DEREGULATION IN THE EUROPEAN UNION

Can deregulation help solve Europe's environmental problems?

In the European Union (EU), the past two decades have seen an immense growth in regulatory measures aimed at environmental protection. In recent years, this regulatory activity has come under increasing criticism. This criticism has coincided with a more general disenchantment with regulation, resulting in a wave of 'deregulation' initiatives. These initially focused on privatisation and market liberalisation in various industries (economic deregulation), but subsequently have also been applied to environmental policy itself. While the term deregulation has become a real 'buzzword', in reality the shift is in regulatory focus rather than deregulation as such.

Deregulation in the European Union looks at two separate but related facets of deregulation. First, it examines the environmental implications of economic deregulation through case studies of the energy, transport and water sectors. Second, it deals with options for deregulation in environmental policy, looking at self-regulation, negotiated agreements and environmental management systems. A number of other issues are also addressed, such as the links between deregulation, environmental protection and competitiveness, and the importance of greater transparency and better availability of environmental information. Presenting evidence from a number of EU Member States and Hungary, one of the most likely contenders for EU membership, this book points at particular challenges facing the countries of Central and Eastern Europe.

Deregulation in the European Union reveals that, from an environmental perspective, deregulation can be both an opportunity and a threat.

Ute Collier is Senior Research Officer, Energy and Climate, at Friends of the Earth Trust, London. Previously, she was Jean Monnet Fellow and Co-ordinator of the Working Group on Environmental Studies at the European University Institute. Florence.

DEREGULATION IN THE EUROPEAN UNION

Environmental Perspectives

Edited by Ute Collier

London and New York

First published 1998
by Routledge
11 New Fetter Lane, London EC4P 4EE

Reprinted 1999

Simultaneously published in the USA and Canada
by Routledge
29 West 35th Street, New York, NY 10001

Editorial matter © 1998 Ute Collier
Individual contributions © 1998 respective contributors
Collection © 1998 Routledge

Routledge is an imprint of the Taylor & Francis Group

Typeset in Garamond by LaserScript Limited, Mitcham, Surrey

Printed and bound in Great Britian by T.J.I. Digital, Padstow, Cornwall

British Library Cataloguing in Publication Data
A catalogue record for this book is available from the British Library

Library of Congress Cataloguing in Publication Data
A catalogue record for this book has been requested

ISBN 0–415–15694–7

CONTENTS

v

CONTENTS

FIGURES AND TABLES

FIGURES

TABLES

CONTRIBUTORS

Gyula Bándi is Professor of Environmental Law at the Eötvös Loránd University, Budapest. He is founder and President of the Hungarian Environmental Management and Law Association.

Ute Collier is Senior Research Officer, Energy and Climate, at Friends of the Earth Trust in London. Until February 1997, she was a Research Fellow at the Robert Schuman Centre, European University Institute, Florence. She has recently co-edited *Cases in Climate Change Policy: Political Reality in the European Union* (with Ragnar Löfstedt, Earthscan, London, 1997) and *Subsidiarity and Shared Responsibility: New Challenges for EU Environmental Policy* (with Jonathan Golub and Alexander Kreher, Nomos Verlag, Baden-Baden, 1996).

Simon Cowan is Wigmore Fellow and Tutor in Economics at Worcester College, Oxford and University Lecturer in Economics at the University of Oxford. He is co-author (with John Vickers and Mark Armstrong) of *Regulatory Reform: Economic Analysis and British Experience* (MIT Press, Massachusetts, 1994).

Wyn Grant is Chair of the Department of Politics and International Studies at the University of Warwick. A recent book is *Autos, Smog and Pollution Control: the Politics of Air Quality Management in California* (Edward Elgar, Cheltenham, 1995).

Veerle Heyvaert is a Researcher in the Department of Law at the European University Institute. She is currently completing her Ph.D. on the regulation of chemical substances in Europe.

Karl-Heinz Ladeur is Professor of Law in the Department of Law, University of Hamburg. He has recently published *Das Umweltrecht der Wissensgesellschaft: von der Gefahrenabwehr zum Risikomanagement* (Dunker & Humblot, Berlin, 1995).

Duncan Liefferink is a Research Fellow at the Department of Sociology, University of Wageningen, the Netherlands. He is author of *Environment* *and the Nation State: the Netherlands, the European Union and Acid Rain* (Manchester University Press, Manchester, 1996).

Arthur P.J. Mol is a Lecturer in the Department of Sociology, University of Wageningen, the Netherlands. He is author of *The Refinement of Production. Ecological Modernisation Theory and the Chemical Industry* (Van Arkel, Utrecht, 1995).

Giorgio Porta is Director for International Activities at ENI SpA. He is Vice-President of CEFIC, the European Federation of Chemical Industries, President of the Italian–Polish Working Group for the Environment and Chairman of UNICE's Environment Working Group.

Richard Starkey is a Researcher at the Centre for Corporate Environmental Management at the University of Huddersfield. He is co-editor (with Richard Welford) of the *Earthscan Reader in Business and the Environment* (Earthscan, London, 1996).

Michael Teutsch is a Researcher in the Department of Social and Political Sciences at the European University Institute. He is participating in a comparative research project examining the impact of European integration on national transport policies.

Frans van der Woerd is Research Fellow at the Institute for Environmental Studies (IVM) of the Vrije Universiteit in Amsterdam. He has recently completed his Ph.D. research on 'Self-regulation in corporate environmental management: changing interactions between companies and authorities'.

ACKNOWLEDGEMENTS

This volume has its roots in a workshop on Deregulation and the Environment, organised by the Working Group on Environmental Studies (WGES) of the European University Institute in Florence in May 1996. The WGES operates under the auspices of the inter-disciplinary Robert Schuman Centre and I was fortunate to be able to act as coordinator of the group from autumn 1995 to spring 1997. The workshop was very much a cooperative effort and many thanks go to those WGES members who helped in making it a success. However, responsibility for this volume (including the selection of specific workshop papers for inclusion) lies solely with the editor.

A particular thanks is due to Professor Yves Mény, Director of the Robert Schuman Centre, for his continuing support of the WGES and its activities. I would also like to thank Professor Mény more generally for supporting my work and encouraging various publication projects during my stay at the European University Institute. Furthermore, I am indebted to Monique Cavallari for administrative and logistical support during the organisation of the workshop.

Thank you also to my friends and colleagues in Florence for providing me with much needed light relief during the editing phase of the book.

Finally, I am grateful to all the chapter authors for responding to my editorial comments and revising their contributions promptly.

Ute Collier, London
May 1997

Part I

DEREGULATION AND LIBERALISATION: NEW CHALLENGES FOR ENVIRONMENTAL POLICY

1

THE ENVIRONMENTAL DIMENSIONS OF DEREGULATION

An Introduction

Ute Collier

INTRODUCTION

Since the mid-1970s, environmental protection has become one of the principal areas of public policy in Europe. Growing awareness about the negative environmental effects of economic activities has resulted in a growth of state intervention, at national, regional and local levels, to achieve environmental objectives. Furthermore, the European Union (EU)'s regulatory activity in this field has expanded enormously. The policy instruments applied to deal with environmental degradation have been wide-ranging, including information campaigns and financial incentives. However, there has been a dominance of so-called command-and-control regulation, with legislation setting a variety of pollution, quality and safety standards. Since the early 1990s, this regulatory activity has come under increasing criticism, in tandem with a more general disenchantment with regulation and government intervention.

Two forces driving the pressure for change can be identified. Firstly, concern about the negative effects of economic regulation on the competitiveness of industry has resulted in a deregulatory drive, focusing on liberalisation of some industrial sectors (in particular the utility industries), as well as the privatisation of state owned companies. The emphasis has been on breaking the powers of monopolies and the introduction of competition. This can have environmental implications in some sectors, such as energy, where competition affects technology and fuel choices.

Secondly, the effectiveness of command-and-control environmental measures has come under scrutiny. As the European Environment Agency (EEA) demonstrated in its 1995 assessment of the state of the environment in Europe, despite some progress towards reducing certain pressures on the environment, this has not been enough to improve environmental quality in general (European Environment Agency, 1995). While the EEA calls for accelerated policies, the question arises as to where exactly the blame for this unsatisfactory situation lies, in view of the greatly expanded regulatory activity in the environmental field. Some would argue that government intervention has not been extensive enough and that standards (and their enforcement) have been too lax. However, another interpretation is that the nature of government intervention is to blame, with command-and-control legislation simply not offering the most effective solution, either in environmental or in cost terms.

Lack of enforcement has been a particular problem, partially for budget reasons. According to Ayres and Braithwaite (1992), governments can simply not afford to do an adequate enforcement job themselves. To amend this situation, there have been calls for the use of different types of instruments, such as environmental taxes (which could also be termed indirect regulation) or voluntary instruments (self-regulation), which work in conjunction with market forces and are thus effectively 'self-enforcing'. Such instruments are also considered more advantageous in terms of industrial competitiveness, resulting in a 'win-win' situation. This compatibility is also at the heart of the sustainable development paradigm (see below) and the theory of ecological modernisation (Hanf, 1994; Weale, 1992).

Although the general regulatory critique has different roots to the criticism of environmental regulation, there are some common assumptions about the preference for the market as a problem-solving tool, with the main task for governments and regulators being to ensure the functioning of market forces. The regulatory changes that have been advocated, be it for competitiveness or for environmental reasons, are often grouped under the umbrella of the term 'deregulation'. Yet, in neither case is it technically correct to speak of deregulation. As Button and Swann (1989) have pointed out, deregulation is too precise a word and in practice encompasses both differing degrees of deregulation and a variety of possible changes in the way in which regulation operates. While deregulation implies the scaling back of the role of the state, liberalisation and privatisation have paradoxically been accompanied by a substantial amount of new regulation, as newly deregulated or privatised industries lose their pre-existing statutory immunity from competition law and other regulatory requirements (Majone, 1994). Also, regulation is often needed to promote competition in areas where there are 'natural' tendencies for concentration.

In the environmental field, state intervention is set to continue, as few advocate no environmental regulation at all. It would thus be more correct to talk about regulatory reform or re-regulation, namely a move from command-and-control instruments to more market-based instruments. According to Hancher and Moran (1989), deregulation is an 'ideological construct' which policy-makers use for a wide range of activities, many of which do not entail the abolition of rules, as the term may imply. In the Netherlands, for example, deregulation has been synonymous with plans for a fundamental reorientation of the relations between state and society (Hanf, 1989). Despite the inexact nature of the term, deregulation has become something of a buzzword in the political debate over recent years. The use of the term in this volume needs to be understood with the above provisos.

The US can be considered the birthplace of the deregulation movement and there, the 1980s have been termed the decade of deregulation (Moran and Prosser, 1994). There has been some spillover into Europe, exemplified by the liberalisation moves in sectors such as telecoms, aviation and the energy sector, with the UK as the European 'pioneer' in these areas. Initially, deregulatory moves focused only on economic regulation, but subsequently, other regulatory areas, including environmental regulation, have come under scrutiny. Furthermore, economic deregulation can have environmental dimensions. However, to date the environmental dimensions of deregulation, and the interlinkages between the two facets of deregulation, have received relatively little attention in the academic literature.

This volume endeavours to contribute to filling this analytical gap, through an emphasis on the links between deregulation moves for economic reasons, and deregulation moves with an environmental rationale. The analysis centres around the assessment of the environmental costs and benefits of these two strands of deregulation, through an interdisciplinary and comparative analysis, with chapters written by experts from a number of disciplines, from a number of European countries. Developments in the EU and a number of its Member States are discussed, with the addition of one chapter discussing the situation in Hungary, one of the strongest contenders for EU membership.

Because of the complexity of the issues involved, this volume cannot provide an exhaustive analysis of the topic of deregulation and the environment. In particular on the theme of environmental deregulation, the reader is also referred to two other volumes in Routledge's EUI Environmental Policy Series. *New Instruments for Environmental Protection in the EU* (Golub, 1997a), provides detailed country case studies (covering Belgium, Germany, the Netherlands, Italy, Spain and the UK), chapters on eco-audits and eco-labels, as well as a further examination of voluntary agreements. *Global Competition and EU External Environmental Policy* (Golub, 1997b) explores how economic competitiveness and regulatory reform have altered the development of the external dimensions of environmental protection in the EU (including a look at international trade, the Common Agricultural Policy, development funds and climate change issues).

The objective of this introductory chapter is to trace the development of the deregulation ideology and to identify some of the environmental implications of deregulation and liberalisation, so as to set the context for the rest of the volume. The chapter commences with an examination of the arguments generally advanced in favour of regulation, both in economic and environmental terms. It then discusses the rationale for deregulation, with a particular focus on establishing the link between deregulation and sustainable development. This is followed by a summary of how EU policy-making has been affected by deregulation moves. Finally, the chapter provides a synthesis of the rest of the volume.

THE NEED FOR REGULATION

In broad terms, regulation is the intervention of the state in private spheres of activity (Francis, 1993). Economists talk of state intervention in the market, the market effectively being a broad range of social activities. According to neo-classical economic models, markets have to be perfect to function, which includes requirements such as full competition, free access and perfect information. In reality, these preconditions are never met in full and markets have numerous failures. Hence, there is a need for intervention to ensure that public interests are being protected. This intervention is most commonly undertaken by governments through a regulatory regime.

Regulatory activities can be divided into two distinct types: economic regulation and social regulation. Economic regulation occurs in response to imperfect competition, and consists of regulations setting price levels (e.g. price cap regulation), limiting profits (rate of return regulation) or specifying certain conditions of operation (conduct regulation). These are generally targeted at private firms, although publicly owned companies can also be regulated in this way and otherwise be left to operate like private firms.

Economic regulation and public ownership

Economic regulation is particularly prevalent where natural monopolies are believed to exist, for example in the utility industries such as water, telecommunications and electricity (Bishop, Kay and Mayer, 1995). Natural monopolies can be characterised by the high threshold costs for participation (Francis, 1993). Because of the capital-intensive infrastructure requirements of these industries, entry into the market by new competitors is restricted 'naturally'. Provision by a single supplier is the most efficient outcome as it avoids wasteful duplication (OECD, 1992). In the past, government intervention in natural monopolies has mainly taken the shape of public ownership. In many cases, publicly owned companies have been under the direct control of government ministries, but have not been subject to statutory requirements such as competition law.

There has often been an assumption that public ownership carries with it the automatic realisation of social and economic regulatory goals (Francis,

1993) and the protection of the public interest (Majone, 1994), thus exempting them from the regulatory regime private monopolies have to comply with. Certainly, it has often been the case that public companies are not expected to yield the high levels of rate of return found in the private sector. Loss-making might be accepted if other public policy objectives, such as employment creation, are achieved. However, as has been pointed out by Majone (1990, 1994), it is questionable whether public companies automatically operate in the best interest of society as a whole. As far as environmental performance is concerned, public companies have often performed no better than private ones, in particular where environmental improvement does not feature amongst the objectives set by governments for the companies. A case in point is the operation of the UK Central Electricity Board. While it played a role in keeping the UK coal industry alive, its efforts in the environmental protection field were rather limited. One problem is that public industries generally come under the control of Industry Ministries, who may act to promote general economic objectives but are not necessarily best placed for realising social and environmental objectives. Nevertheless, in some cases, public ownership has served as a vehicle to promote environmental aims, as the case of local energy companies in Germany demonstrates (see Chapter 6). Overall though, environmental aims have generally been addressed through specific regulatory measures.

Environmental regulation

The need for environmental regulation, which forms one of the sub-areas of social regulation, stems from the fact that there are market failures related to public goods, such as air and water. As these public goods are not priced, they are not taken into account in private production and consumption decisions and so-called external costs occur. Furthermore, sometimes it is government intervention itself which can create environmental problems. A good example here is the Common Agricultural Policy, which has operated to meet various social and economic objectives, but has resulted in activities which are clearly environmentally damaging (Lenschow, 1997). The same is true for various policy decisions in other areas, such as energy and transport policy. Here, again, the environmental costs are largely external to the costs to be borne by decision-makers and hence fail to be taken into account.

Environmental regulation can take a variety of forms, depending on the type and scale of the environmental issue to be addressed. Some problems have lent themselves to command-and control legislation, for example in the case of emissions of air and water from certain industrial plants. Often, strict standards have been set, which are relatively straightforward to enforce but can be very costly to industry. However, problem-solving can become much more complex, for example where the assignment of causal

7

and legal responsibility is difficult (Francis, 1993). To deal with different types of problems, environmental policy has always consisted of a range of instruments, including information campaigns and financial subsidies. Furthermore, various levels of government have been involved in the design and implementation of these instruments, with an increasing involvement at the supra-national level. The EU has emerged as an important regulatory force, with a focus on command-and-control style Directives (Collier, 1996a).

While environmental regulation has seen a substantial growth over recent decades, it has not been without its critics. Francis (1993) argues that social regulatory regimes are generally subject to manifold tensions which affect their continuing viability. The environmental area is particularly vulnerable, as it is substantially affected by value conflicts and scientific uncertainties. These tensions tend to become very prominent during times of economic recession, with criticism abounding about the undue economic burdens of environmental regulations. Command-and-control regulations are particularly subject to attack for their often high costs of compliance and the lack of flexibility they offer industry, which has been accused of hindering technological innovations. However, there have also been criticisms about certain types of regulations from an environmental point of view. Standards can be too inflexible to promote innovations which may have greater environmental benefits. There can be high administration costs and companies may move to countries with lower standards (von Weizsäcker, 1990), effectively resulting in the export of environmental damage as a consequence of regulation. Furthermore, enforcement has been problematic in many cases (according to Hanf, 1989, the 'weakest link in the regulatory chain'), making any approaches relying on market forces and self-regulation an attractive proposition. Hence, both economic and environmental arguments have been advanced for deregulation and regulatory change.

THE ECONOMIC RATIONALE FOR DEREGULATION

The interest in deregulation can be traced back to the early 1960s, when a number of academic economists, especially in the US, began to produce a body of literature highly critical of price, entry and exit regulation (Derthick and Quirk, 1985). One of the criticisms was aimed at the regulators themselves, implying that they had been 'captured' by the regulated industries and were no longer serving the public interest (Francis, 1993; Majone, 1996). The arguments of the research eventually filtered through to government level during the economic recession of the 1970s. Deregulation subsequently became the prevailing policy fashion, first in the US but soon followed by a number of EU countries. It was initially firmly directed at economic regulation in sectors such as telecommunications, transport and

8

various public utilities, with limited effect on other regulatory areas (Noll, 1991).

For these sectors, it had become increasingly evident that public sector regulation, particularly through direct ownership, has been largely ineffective in reducing market failures (Bishop, Kay and Mayer, 1995). Furthermore, the natural monopoly rationale has lost some of its currency in recent years. In the electricity sector, for example, technological developments mean that there are now generating options which can operate economically at a smaller scale. In the telecoms sector, the advent of fibre optics and other technical innovations have resulted in much lower costs for entry into the market. As Hancher and Moran (1989) have observed, in such cases deregulation occurs as existing rules no longer fulfil their original purpose. However, some natural monopoly elements still exist, for example in electricity grids or gas networks. Yet, even in these areas, deregulation has taken a foothold. The approach here has been to create, in some ways artificially, conditions for competition. To maintain these conditions has become the primary task of newly created regulatory agencies, through measures such as liberalising access to grids and networks, breaking up of monopolies, and price regulation.

While there have been some technological factors facilitating the move towards deregulation and liberalisation, there can be little doubt about the importance of a major ideological shift concerning the role of the state in the economy (Francis, 1993; Swann, 1988). Initially, this neo-liberal doctrine was only supported by parties on the right of the political spectrum, but it has subsequently increasingly found favour with parties on the left as well. While previously, state regulatory regimes were seen as crucial in protecting the public from the 'predations' of monopolies (Francis, 1993), now they are often seen as obstructing economic efficiency and the operation of market forces.

The pursuit of economic efficiency has become an ever more salient issue in view of increasing economic globalisation and the resulting perceived threat to the competitiveness of European firms. Inefficient firms imply lower profits, higher prices, lack of competitiveness and unemployment. The link between regulation and inefficiency is far from proven but has nevertheless received substantial attention in recent years. For example, a report by the Economists Advisory Group to the European Commission (1987) identified four sources of inefficiency due to regulation:

- technical, due to the protection of inefficiently sized firms;
- internal, due to the lack of competitive pressures;
- allocative, as regulation tolerates a price structure out of line with costs, above the level that would prevail under competition;
- innovative, due to the absence of competition and regulatory restrictions on cross-industry diversification.

The proponents of competition argue that it creates an incentive mechanism for more efficient operation, as well as opportunities for innovation. Paradoxically, as already mentioned, liberalisation in a number of sectors has meant an increase of regulation, specifically to promote competition. However, regulatory activities in this context have often been passed to independent or semi-autonomous regulatory agencies, who are assumed to be less prone to political influence and more expert in their actions.

The relationship between competition, efficiency and regulation is thus clearly complex but it is beyond the scope of this volume to enter this discussion in any greater depth. Instead, the issue of interest here is the identification of the environmental costs and benefits of market liberalisation. These have received little attention in the literature and the deregulation drive has been conducted almost exclusively on the basis of economic efficiency arguments. This rationale rests on a very narrow definition of efficiency, focusing on the minimisation of operational costs and the maximisation of profits. As Vickers and Yarrow (1988) have argued, there are often trade-offs between welfare maximisation and profit maximisation in a more competitive market. Some of these trade-offs are likely to be at the expense of environmental protection.

THE ENVIRONMENTAL DIMENSION

Superficially, liberalisation and economic deregulation appear to have little relevance for environmental protection. However, some of the sectors which have been the focus of deregulatory moves, namely, energy, transport and water, are responsible for a considerable amount of environmental damage. Where there was public ownership of these industries, it was often assumed that their operation was to meet a range of government policy objectives, generally in the social field (e.g. employment creation). However, as already mentioned, there are examples of public sector companies making environmentally beneficial investments, even though they are not economically beneficial or a regulatory requirement. Such investments cannot necessarily be expected from private companies operating in a liberalised market. Furthermore, liberalisation itself might cause increased environmental damage, for example through associated increases in transport volumes (see Chapter 7). Hence, the establishment of a regulatory regime is of crucial importance to ensure an environmentally responsible behaviour by companies and to minimise additional environmental damage. Deregulation and liberalisation have been accompanied by the establishment of new regulatory agencies, but their remit does not necessarily give sufficient consideration to environmental concerns.

Furthermore, in some cases (e.g. the promotion of energy efficiency in UK energy sector privatisation), there has been a presumption that competition itself will serve to achieve environmental objectives, just as it

10

once was assumed that public ownership could be a vehicle for achieving social objectives. As already mentioned, the existence of externalities means that the market itself should not be relied on for the best outcome in environmental terms. This does not mean that there are no opportunities for using certain market features to make environmental protection more effective. It is within this context that the deregulation doctrine has affected environmental regulation itself and has been closely associated with the sustainable development debate. To date, the underlying rationale of environmental protection was never questioned seriously and the issue under discussion is not whether to deregulate but rather how, and how extensively to regulate. In the Netherlands, generally considered one of the lead countries in environmental policy in the EU, environmental deregulation became a big theme in the early 1980s, with a Deregulation Action Programme adopted in 1983. However, as Hanf (1989) has discussed, rather than abolishing regulation, this has involved the retention of the overall regulatory objectives and aimed at developing alternatives to the more traditional instruments of regulation.

An important part of the Dutch approach has been the stimulation of target groups to take appropriate measures to promote environmental quality. Within this context, deregulation is an attempt to influence the behaviour of environmentally relevant actors in a more effective and, particularly, more efficient way (Hanf, 1994). He argues that:

> Deregulation is less about leaving environmental quality at the mercy of free market forces and more about the relation between the instruments to be used in pursuing these objectives and the impact of these policy constraints on the ability of the affected firm to act efficiently in the market place.

The idea is that policy instruments should inhibit as little as possible the 'market efficient behaviour' of economic actors. Hanf suggests that in the Netherlands deregulation has not fundamentally altered the regulatory constraints on economic activity, but has led to a restructuring of the regulatory space around a point of equilibrium between concern for environmental quality on the part of economic actors and improved economic competitiveness of firms as a result of increased responsiveness to market forces. This kind of environmental 'deregulation' is part and parcel of the sustainable development paradigm which has gained currency since the early 1990s.

LINKING DEREGULATION TO SUSTAINABLE DEVELOPMENT

An important development in recent years has been the growing realisation that environmental problems cannot necessarily be solved through 'end-of-pipe' solutions. As already mentioned, while market failures are one cause of environmental damage, it can also result from government intervention in other policy areas, such as agriculture or energy. There is thus a need for change in these policy areas, towards a greater compatibility with environmental objectives. The concept of policy integration (i.e. the integration of environmental concerns into other policy areas) has received increased attention in recent years. This has gone hand in hand with the realisation that current patterns of economic development cannot be sustained indefinitely, neither in environmental (resource use, damage) nor in social terms (unemployment). Sustainable development has thus become, at least in principle, one of the major guiding principles of both government and industrial activities.

As has been discussed at length elsewhere, the sustainable development concept is subject to a variety of interpretations (see e.g. Common, 1995; Jacobs, 1991). However, the different interpretations share the common assumption that economic development and environmental protection are interdependent and that they need to be made compatible. The EU's Fifth Environmental Action Programme (EAP), for example, defines 'sustainable' as reflecting:

> a policy and strategy for continued economic and social development without detriment to the environment and the natural resources on the quality of which continued human activity and further development depend.
>
> (European Commission, 1992)

The mainstays of a shift to sustainable development are believed to be policy actions which make both economic and environmental sense, so-called 'no-regrets' options (also frequently referred to as 'win-win' situations, or the 'double dividend'). An obvious example here would be energy efficiency improvements, of which there are many cost-effective examples. Yet, because of the existence of a variety of market failures, such improvements are currently not being made. The solution is thus assumed to be a policy which is based on remedying such market failures, especially those, such as taxes and charges, which aim at internalising the external costs of production (Collier and Löfstedt, 1997). A market-based approach is also supposed to leave greater flexibility for economic actors in achieving environmental objectives.

The use of taxes and charges, as a means of internalising the clear external costs associated with various economic activities, appears promising. However, the valuation of these external costs is far from straightforward. There are clear methodological problems and ethical questions which arise about the intrinsic value of nature, as well as of human life (Common, 1995; see also Chapter 2 in this volume). Jacobs (1991) argues that valuation is not strictly necessary for applying environmental taxes, as long as the relationship between demand and price is known. In such a pragmatic solution, the level of tax would be determined by the amount of demand to be reduced. However, in the case of a carbon tax, the low price elasticity of energy demand, as well as a range of other market failures mean that to be effective, a carbon tax would probably have to be set at a level which would be highly socially regressive, economically damaging and politically unacceptable. Furthermore, there are suggestions that taxes are less reliable than regulatory tools (OECD, 1993) in that polluters may choose to pay the tax and continue to pollute. However, non-compliance is also a serious problem with some command-and-control regulations and only experience will tell which instrument has a better performance.

As taxes and charges are covered in some detail in the companion volume on new instruments (Golub, 1997a), the chapters of this volume (in particular those of Part III) focus on instruments which require a more active role from business and industry, which also fits squarely with the deregulation ideology. The EU's Fifth EAP stresses the need for 'shared responsibility', implying the use of a diversity of instruments and the involvement of all relevant actors in the policy process (European Commission, 1992). This includes an enhanced role for 'self-regulation' and 'soft' instruments, such as voluntary (or negotiated) agreements with industry, environmental management systems and labelling. Such instruments appear attractive as they involve no (or very small) administrative costs for public bodies and should be 'self-enforcing'. In this context the provision of environmental information becomes ever more important, so that consumers and pressure groups can exert pressure on companies to ensure compliance (see Chapter 4).

DEREGULATION INITIATIVES AT THE EU LEVEL

In the EU, deregulatory activities have three separate, but interlinked dimensions. Firstly, the deregulation philosophy lies at the heart of the efforts to complete the Single European Market (SEM). The SEM aims at the liberalisation of monopolised and heavily regulated markets, so as to strengthen the competitiveness of European industry. The 1992 programme for the completion of the SEM focused on the removal of obstacles in a number of areas, including the simplification of some regulations. One of the most recent achievements has been the agreement on the Internal Energy Market, with Member States gradually opening up their previously protected energy markets. The Single Market initiative paid little attention to the potential environmental implications of liberalisation, although a report on the Single Market and the Environment was published in 1990, two years after agreement on the completion of the SEM. The report provided a rather inconclusive assessment, suggesting that in environmental terms, the SEM presented both challenges and opportunities (Task Force Environment and Internal Market, 1990). The identification and quantification of the environmental effects of pan-European liberalisation is far from straightforward, but doubts have to be raised about the compatibility between the ever increasing volumes of production and trade (and associated transport volumes), and objectives such as resource conservation.

Secondly, the increasing questioning of supra-national intervention at the expense of national sovereignty can also be viewed in deregulation terms. The subsidiarity concept has become increasingly influential since the signing of the Treaty on European Union. While a repatriation of responsibilities to the national level does not necessarily have to mean deregulation, in some cases national Member States will see this as a welcome opportunity to loosen regulatory intervention. It is beyond the scope of this volume to explore this issue any further, and a more in-depth assessment can be found in Collier, Golub and Kreher (1996).

Parallel to these developments, the sustainable development concept has begun to influence EU environmental policy. It is reflected in the theme of the Fifth EAP (entitled 'Towards Sustainability'), and the ratification of the Treaty of European Union saw the inclusion of 'sustainable growth' as one of the main tasks of the Union. While generally somewhat vague about the exact means and targets to achieve a transition to a sustainable future, the Fifth EAP suggests that 'valuations, pricing and accounting mechanisms have a pivotal role to play in the achievement of sustainable development'. Furthermore, 'economic and fiscal instruments will have to constitute an increasingly important part of the overall approach' (to environmental policy) (European Commission, 1992).

Some discussions have taken place about the possibility of integrating greater competitiveness and environmental protection, most notably in the 1993 White Paper on Growth, Competitiveness and Employment (European Commission, 1993). The basis of this was a broad tax reform, which included the levying of environmental and energy taxes, with concurrent reductions in labour taxes. The paper stressed the promise of the so-called double dividend, suggesting that environmental damage could be decreased, while employment increased. However, this link is far from proven and the idea has since disappeared from the agenda.

Meanwhile, attempts to introduce an EU level carbon/energy tax have failed on account of subsidiarity considerations (Collier, 1996b). The experience with economic instruments is thus restricted to the Member States, a number of which have introduced carbon taxes, as well as some other environmental taxes unilaterally. Some useful experience is being gained, which is discussed in more detail in Golub (1997a).

Other types of market-based approaches have also received increasing attention. At EU level, regulations on eco-labelling and on eco-auditing are in place. Neither of these are compulsory, which renders their effectiveness rather uncertain. As far as voluntary or negotiated agreements are concerned, there is no concrete experience with EU wide agreements. In June 1996, the Council gave the Commission a mandate to negotiate a voluntary agreement with car manufacturers on fuel efficiency, as a means to reduce CO_2 emissions from the transport sector. Previously, there had been discussion about mandatory standards and it is not clear how effective such voluntary agreement, if reached, would be, considering the rather equivocal experience in the Netherlands (see Chapter 11).

Despite these developments, the bulk of EU environmental protection instruments is of the command-and-control type. However, this approach is increasingly coming under pressure. In September 1994, the Commission established a high-level group of 'independent experts' to examine the impact of EU and national legislation on employment and competitiveness. The Molitor Report, named after the chairman of the group, was published in 1995 and chose environmental legislation as one of only four sectors examined. The report advocates a new approach to environmental regulation:

> which stresses the setting of general environmental targets whilst leaving the Member States and, in particular, industry the flexibility to choose the means of implementation.
>
> (European Commission, 1995)

This echoes the calls for both subsidiarity and deregulation. At the same time, the group also advocates the use of market-based instruments:

> Any new proposal should be accompanied by a careful analysis whether or not market-based methods could be employed to achieve the same goals; where a market-based approach is feasible, any departures from it should be justified.
>
> (European Commission, 1995)

The report received some criticism for its methodology and narrow focus. According to the dissenting opinions of one of the members of the group, the report treats environmental protection as an obstacle to economic aims, rather than understanding the interdependence between economy and environment. The report may have been somewhat simplistic in its assessment, and hence not overly influential, but pressure for regulatory changes in environmental policy is likely to continue, both at EU level and in the Member States. At the same time, deregulation and liberalisation moves are continuing in various industrial sectors, in a number of EU Member States, as well as in Eastern Europe. In both cases, economic and political issues dominate the discussion and the assessment of the environmental implications of deregulation has been rather limited. The aim of this volume is to contribute to this assessment through the exploration of a number of different aspects of deregulation.

OUTLINE OF THE CHAPTERS

Part I

Part I of this volume focuses on a number of general issues associated with the application of deregulation in environmental policy. Richard Starkey discusses some fundamental questions that need to be addressed when looking for opportunities to deregulate environmental policy. He attempts to assess to what extent the environment should be protected, using the concept of obligations to future generations. The gist of his argument is that in the search for double dividend and the preoccupation with competitiveness, we have lost sight of the real issue at stake in sustainability terms, namely, the need to maintain environmental carrying capacity.

Within this context, Starkey examines the appropriate roles for business and government. While business can be seen to have become more environmentally aware, he argues against expecting too much from self-regulation. Although such activities are worthwhile, they cannot be sufficient to guarantee the maintenance of environmental carrying capacity. It has to be the responsibility of the state to elaborate targets and to implement appropriate policies and measures. Starkey also considers the arguments advocated by some analysts that competitiveness can be enhanced by the implementation of strict environmental regulation. Here the evidence is inconclusive but Starkey argues that we must accept that there may be costs for environmental protection.

Karl-Heinz Ladeur discusses deregulation within the context of knowledge generation, through the example of safety controls. This is one of the crucial areas in the context of industrial competitiveness discussions, as standards which are too strict can clearly have negative consequences for industrial innovation. Regulations in this area have been subject to continual change as the knowledge base has expanded and the regulatory system has become increasingly differentiated. However, the system has been ill adapted to cope with technologies which have long-term effects (e.g. nuclear power) or are difficult to manage due to insufficient knowledge (e.g. biotechnology). Ladeur thus argues that there is now a need for a comprehensive remodelling of the entire network of relationships between technology and safety control, the knowledge base and normative framework, and private organisation and public administrative processes.

He argues not for deregulation, but for reduced and simplified adaptive regulation, which should foster a system design of technological processes which include the problems of safety control and monitoring under the aspect of risk information and self-revision. The idea would be to replace rule-based regulation with a second-order form of procedural regulation in the sense of adopting a comprehensive reciprocal learning strategy as the

16

basis for a new coordination between industry and administrative decision-making. This approach does not rule out the use of economic instruments in some domains, but Ladeur warns of misplaced incentives and short-sighted trade-offs which may suffocate the development of promising technologies.

While Ladeur points to the increasing complexity of the knowledge base, one of the other problems of the traditional regulatory system is the lack of access to this information by actors outside the regulatory process, such as pressure groups and consumers. Veerle Heyvaert argues that if deregulation is to become a reality, a movement to greater transparency and availability of environmental information is an indispensable mechanism to correct some of the shortcomings of deregulation. Because of its awareness-heightening and mobilising effects on consumers, access to information contributes to correcting the market failure caused by a lack of internalisation of costs.

Some initial steps towards better access have been made through the EU Freedom of Access to Information on the Environment Directive, and the endorsement to access in Agenda 21, as well as a draft UN/ECE convention on this issue. In environmental policy, deregulation must inevitably deal with the paradox that it hands over the care of the environment to those parties (i.e. industry) which traditionally – and often deservedly – have been labelled its enemies. Thus, whereas deregulation leads to more internalised forms of environmental management, it is highly doubtful whether it will entail a reduced call for external control. Furthermore, questions arise as to how commercially sensitive data is to be dealt with. This issue will become more prevalent as the business environment becomes ever more competitive. Some level of involvement of public authorities is thus clearly necessary.

Access to information is one of the principles included in the new environmental protection acts that have recently been adopted in the countries of Central and Eastern Europe. In these countries, there is an urgent need to reform the legal system and to establish an efficient system for environmental protection. In the past, laws were wide-ranging but compliance poor. Gyula Bándi focuses on the example of Hungary to examine the challenge of integrating the need for environmental regulation into the framework of a developing market economy. He stresses that while there is some potential for market based instruments, a certain amount of command-and-control legislation is indispensable.

The new Hungarian Environmental Protection Act makes various provisions for environmental management, as well as public participation, two instruments relevant in the deregulation context. Bándi argues that in both cases, there is a requirement for a detailed set of guarantees and guidelines, which need to be set by legislation. As far as public participation is concerned, currently in the CEE countries there are numerous drafts of such legislation but no clear procedures have been established to date. As

17

far as environmental management is concerned, provisions for eco-auditing have been included in the Hungarian environmental act and there is hope that such systems may contribute to companies meeting environmental requirements.

Part II

The second part of this volume focuses on sectoral perspectives. The chapters analyse three of the sectors which have been the focus of economic deregulation moves: energy, transport and water. Ute Collier examines the case of energy sector liberalisation, arguing that it can in principle have both positive and negative environmental implications. It promises a great opportunity for shaking up a monopolistic energy market, dominated by supply-oriented companies, focused on large-scale generation technologies, especially coal and nuclear. However, a 'free for all' situation could mean damaging competition between options such as gas versus district heating or combined cycle gas plants versus renewables. The best outcome in environmental terms cannot be assumed, especially while multiple market failures exist, including those related to external costs.

Collier analyses the liberalisation issue by looking at two contrasting cases: the privatised, liberalised electricity sector in the UK and the monopolistic, publicly owned local energy companies in Germany. Her argument is that a liberalised system, while providing some environmental benefits, requires substantial regulatory intervention, especially in the area of energy efficiency. At the same time, substantial environmental benefits can be realised under public ownership, especially at the local level, provided the political climate is favourable, as in Germany. For this system, liberalisation is a considerable threat and should not necessarily be pursued at all cost.

Michael Teutsch reaches similarly mixed conclusions about liberalisation and deregulation in the transport sector. The liberalisation of transport operations, such as road haulage, does not necessarily have negative environmental implications per se. Past interventions such as quantitative restrictions for market access and price-setting have not been able to guarantee a sensible allocation of goods, neither in economic nor in environmental terms. Deregulation might present an opportunity for improving the situation.

Within this context, Teutsch discusses the importance of correct pricing for reducing the environmental damage caused by the transport sector. Liberalisation thus needs to be accompanied by taxes and charges to ensure greater environmental compatibility. This has been recognised in the EU, as a growing number of Commission statements on this issue demonstrate. Environmental cost internalisation also needs to be accompanied by other measures, making the most environmentally benign transport options (railways, inland waterways) more attractive. However, in both cases there is a great reluctance by the Member States to agree on common measures, while at the national level little progress is evident.

The water sector has also seen privatisation and liberalisation in a number of countries. Dealing with water pollution involves both controlling the operation of private water companies, as well as a large number of industrial production sites. Simon Cowan assesses the role of alternative instruments for water pollution from an economic perspective. His analysis suggests that environmental problems in the water industry are unlikely to be solved by voluntary instruments and civil liability. He does see a role for instruments such as environmental charges but in addition to, rather than instead of, quantitative regulations.

In terms of the environmental effects of liberalisation, Cowan draws on the UK experience. Before privatisation, the public water boards were effectively regulating themselves and environmental protection objectives were not well served. Another problem was the unwillingness of government to provide funds for a water quality investment programme. Privatisation has meant that the enforcement of legislation is easier, the transparency of environmental objectives is more obvious and there has been a boom in investment, with most of the money being spent on improvements in sewage treatment and disposal.

Part III

Part III focuses on how industry performs in a deregulatory climate. Wyn Grant argues that it is necessary to make a distinction between large firms and small and medium-sized enterprises (SMEs) when discussing the implications of regulation and deregulation. Generally, large firms have greater opportunities to influence the agenda of environmental regulation (or deregulation), which does have implications for competitiveness and market structures. In order to understand the move towards environmental deregulation, Grant discusses developments in the US, where in March 1995 President Clinton launched the Reinventing Environmental Regulation Initiative, designed to encourage collaboration between industry and regulators. In Europe, the debate has centred on the Molitor Report and also on the UNICE Regulatory Report entitled 'Releasing Europe's Potential Through Targeted Regulatory Reform'. While this report focuses on SMEs, Grant is not convinced that its proposals will be implemented, as long as big business holds the balance of power.

Giorgio Porta is chairman of UNICE's environmental working group, which was responsible for the 1995 report on regulatory reform. The report calls for simplification and re-orientation of regulation, higher quality regulation, both at national and at EU level. It does not advocate de-regulation per se, but pleads for it in some cases. UNICE is very much in favour of negotiated agreements (mainly at national/regional level) as a way to ensure more efficient environment policy measures. Porta argues that they have a number of advantages. Both parties can rely on a common ground for measuring the issue at stake, setting clear objectives and identifying possible solutions. He maintains that industry feels more committed to fulfilling objectives it has contributed to defining. However, he concedes that control mechanisms are essential for the credibility of agreements.

The Netherlands is generally considered as a lead country in the area of negotiated agreements and Duncan Liefferink and Arthur Mol examine the recent experience with these instruments. The first part of the chapter discusses the emergence of voluntary agreements in the broader context of shifting relations between state and society. It argues that in the environmental field, the shift towards more cooperative and communicative approaches can also be seen as an aspect of the process of ecological modernisation in the political realm. Liefferink and Mol then review the rich experience gained with voluntary agreements in the Netherlands. Currently, more than 100 voluntary agreements exist, covering almost all aspects of environmental policy. The case of the dairy industry is discussed in more detail, looking at the energy efficiency and packaging covenants. Liefferink and Mol's conclusion is that the status of environmental covenants is still unclear and controversial, in particular as far as questions of enforcement are concerned.

Frans van der Woerd focuses on the Dutch experience with Environmental Management Systems (EMS), which he considers as a complement to voluntary agreements. Between 1989 and 1995, EMS saw a steady development in large Dutch companies, in response to increasing calls for self-regulation. Company licensing procedures have been increasingly linked to the implementation of such EMS systems, constituting a form of deregulation. Van der Woerd discusses case studies of four companies, which identify some opportunities but also bottlenecks. He suggests that the flow of information that is being generated is actually more important than the formal permits. There are possibilities for win-win solutions but no free lunch.

Finally, a short epilogue draws some conclusions about the future of environmental deregulation in the EU.

REFERENCES

Ayres, I. and Braithwaite, J. (1992) *Responsive Regulation: Transcending the Deregulation Debate*, Oxford: Oxford University Press.

Bishop, M., Kay, J. and Mayer, C. (1995) *The Regulatory Challenge*, Oxford: Oxford University Press.

Button, K. and Swann, D. (1989) *The Age of Regulatory Reform*, Oxford: Clarendon Press.

Christensen, J.O. (1989) 'Regulation, deregulation and public democracy', *European Journal of Political Research* 17, pp. 223–239.

Collier, U. (1996a) *Deregulation, Subsidiarity and Sustainability: New Challenges for European Union Environmental Policy*, Working Paper, Robert Schuman Centre, European University Institute.

—— (1996b) 'The European Union's climate change policy: limiting emissions or limiting powers?', *Journal of European Public Policy*, 3 (1), pp. 123–139.

Collier, U., Golub, J. and Kreher, A. (eds) (1996) *Subsidiarity and Shared Responsibility: New Challenges for EU Environmental Policy*, Baden-Baden: Nomos Verlag.

Collier, U. and Löfstedt, R.E. (1997) *Cases in Climate Change Policy: Political Reality in the European Union*, London: Earthscan.

Common, M. (1995) *Sustainability and Policy*, Cambridge: Cambridge University Press.

Derthick, M. and Quirk, P.J. (1985) *The Politics of Deregulation*, Washington: Brookings Institute.

European Commission (1987) *The Likely Impact of Deregulation on Industrial Structures and Competition in the Community*, Luxembourg: Office for Official Publications.

—— (1992), 'Towards sustainability', *COM* (92) 23 final.

—— (1993) 'Growth, competitiveness and employment: the challenges and ways forward into the 21st century', *COM* (93) 700 final.

—— (1995) 'Report of the group of independent experts on legislative and administrative simplification', *COM* (95) 288 final.

European Environment Agency (1995) *Europe's Environment: the Dobris Assessment*, London: Earthscan.

Francis, J. (1993) *The Politics of Regulation*, Oxford/Cambridge: Blackwell.

Golub, J. (ed.) (1997a) *New Instruments for Environmental Protection in the EU*, London: Routledge.

—— (ed.) (1997b) *Global Competition and EU External Environmental Policy*, London: Routledge.

Hancher, L. and Moran, M. (1989) 'Introduction: regulation and deregulation', *European Journal of Political Research* 17, pp. 129–136.

Hanf, K. (1989) 'Deregulation as regulatory reform: the case of environmental policy in the Netherlands', *European Journal of Political Research* 17, pp. 193–207.

—— (1994) 'The political economy of ecological modernisation: creating a regulated market for environmental quality', in Moran and Prosser (eds) op. cit.

Jacobs, M. (1991) *The Green Economy*, London: Pluto Press.

Lenschow, A. (1997) 'The world trade dimension of "greening" the EU's common agricultural policy', in Golub, J. (ed.) (1977b).

Majone, G. (1990) *Deregulation or Re-regulation? Regulatory Reform in Europe and the US*, London: Pinter Publishers.

—— (1994) 'Paradoxes of privatisation and deregulation', *Journal of European Public Policy* 1 (1), pp. 53–69.

—— (1996) *Regulating Europe*, London: Routledge.

Moran, M. and Prosser, T. (eds) (1994) *Privatisation and Regulatory Change in Europe*, Buckingham: Open University Press.

Noll, R.G. (1991) *The Economics and Politics of Deregulation*, Jean Monnet Paper, Florence: European University Institute.

OECD (1992) *Regulatory Reform, Privatisation and Competition Policy*, Paris: OECD.

—— (1993) *International Economic Instruments and Climate Change*, Paris: OECD.

Swann, D. (1988) *The Retreat of the State: Deregulation and Privatisation in the UK and US*, New York: Harvester Wheatsheaf.

Task Force Environment and Internal Market (1990) *1992 – the Environmental Dimension of 1992*, Bonn: Economica Verlag.

Vickers, J. and Yarrow, G. (1988) *Privatisation – an Economic Analysis*, Cambridge: The MIT Press.

Von Weizsäcker, E.U. (1990) 'Regulatory reform and the environment: the cause for environmental taxes', in Majone (ed.) (1990).

Weale, A. (1992) *The New Politics of Pollution*, Manchester: Manchester University Press.

2

COMPETITIVENESS, DEREGULATION AND ENVIRONMENTAL PROTECTION

Richard Starkey

INTRODUCTION

In recent years, the costs and benefits of environmental protection measures have come under increased scrutiny, in particular in relation to their impact on industrial competitiveness. Deregulation of environmental policy has been suggested by some as one of the ways to increase competitiveness, while at the same time others have argued to the contrary, namely, that environmental regulation actually enhances competitiveness.

The aim of this chapter is to examine the relationship between competitiveness, deregulation and environmental protection. It starts by arguing that in order to establish a coherent regime for environmental protection it is necessary first to establish the level of protection required and then to select the most appropriate tools to bring this level of protection about. In short it is necessary to establish appropriate ends and means. So what is the appropriate level of environmental protection? This is the question that the first section of the chapter attempts to answer. It is argued that future generations are of equal moral standing to our own and are therefore entitled to inherit an environment that is equal in quality to the environment of today. Or to put it another way, our current generation is under an obligation to preserve environmental carrying capacity over time. Section 1 then goes on to examine preliminary research into the reductions in the use of natural resources that will be necessary to ensure that this obligation is met.

Having looked at ends in the first part of the chapter, the second and third parts go on to look at means. The second part of the chapter examines whether the pursuit of profit by business is, of itself, sufficient to bring about the changes necessary to maintain environmental carrying capacity. It concludes that it is insufficient and argues that some form of market intervention is needed to ensure carrying capacity is maintained. It is argued that intervention in the form of self-regulation by the business community

23

will not bring about the level of environmental protection required and that government has to intervene in the market to ensure that carrying capacity is maintained.

Given this need for government intervention, the third part of the chapter examines current EU environmental policy and argues that tougher intervention than exists at present is required if our obligations to future generations are to be met. The chapter then goes on to examine whether there is a case for deregulation in the light of this need for tougher intervention. Finally the implications of more stringent government intervention for the competitiveness of business within the EU are examined.

1 WHAT IS THE APPROPRIATE LEVEL OF ENVIRONMENTAL PROTECTION?

The inappropriateness of cost–benefit analysis

The tool advocated by economists as the means of establishing the appropriate level of environmental protection is cost–benefit analysis (CBA). CBA is fundamental to the arguments used in the deregulation debate. Under a CBA approach the environment is protected insofar as the benefits of doing so exceed the costs. Or to put it in economic terms: the optimal level of environmental protection occurs when the marginal cost of protection equates with its marginal benefit.

In order to be able to compare the benefits of environmental protection with the costs, it is necessary that both be measured in the same units. Whilst in theory any unit of measurement could be used (Layard and Glaister, 1994), in practice this means that both be given a monetary value. Whilst it is fairly straightforward to measure the costs of environmental protection it is more problematic to put a monetary value on the benefits, i.e. to measure how much people value the environment. The methods used by economists to give the environment a monetary value have been heavily criticised (see, for example, Jacobs, 1991; O'Neill, 1993; Adams, 1995), but as Jacobs (1991) points out:

> the ability of the methods to overcome the problems of measurement is not the primary issue. For in one sense these objections do not dent the [cost–benefit] argument. The environmental economist could acknowledge that it is very difficult to put a value on the environment, and that in practice this approach is highly imperfect. But he or she could argue nevertheless that its conceptual basis is correct. That is, whether or not it can actually be undertaken as the theory suggests, the rational way of understanding environmental protection is still as a trading off of costs and benefits . . . In this sense the problems of measurement are important but they do not overturn the basic principle.

24

However the conceptual basis of CBA has also been questioned. One of the main criticisms levelled at CBA is that it cannot take into account the valuations of future persons, persons not yet born but who will be affected by the outcome of current decisions made regarding the environment and whose valuations are therefore relevant. CBA is based on the notion that the environment can be valued 'but a significant portion of the environment's value cannot be known, since it comes from people excluded from the valuation process' (Jacobs, 1991).

Whilst it is undoubtedly true that people today place value on preserving the environment for future use, these valuations would act as a proxy for the valuations of future persons only if the value placed on the future environment by current generations was equal to the value which would be placed on this future environment by those future generations actually living in it, i.e. if there was no discounting of the future value of the environment. As this cannot be guaranteed (in fact discounting is highly likely), CBA cannot reliably value the environment. As O'Neill (1993) argues:

> while sophisticated versions of environmental economics show that the values and interests of . . . future generations can be incorporated into CBA, they cannot give them proper weight. This failure is particularly evident in the role of social discounting in CBA in which future goods and harms are valued at less than those of the present.

So far it has been argued that CBA cannot fully take into account the value placed on the environment by future generations. The question that needs to be asked is does this matter? Is it acceptable to discount the lives of future generations, or are future generations morally equivalent to our own? According to Jacobs (1991), questions must be answered about the importance of the lives of future people.

Many argue that it is unacceptable to discount the lives of future generations. They believe that the environment should be protected to such a degree that future generations have access to an environment which is of the same quality as our current environment, i.e. that environmental carrying capacity be maintained over time. For instance Costanza and Daly (1992) have stated:

> An important motivation behind the sustainable development discussion is that of a just bequest to future generations. Utility cannot be bequeathed but natural capital can be. Whether future generations use the natural capital we bequeath to them in ways which lead to happiness or misery is beyond our control. We are not responsible for their happiness or utility – only for conserving for them the natural capital that can provide happiness if used wisely.
>
> (Costanza and Daly, 1992)

25

Various justifications have been given for this position. Jacobs (1991) argues that this relationship between generations is the result of 'the Kantian injunction that each generation should do as it would be done by'. He uses a thought experiment to illustrate, asking: if we were to imagine that we were living in one hundred years time, what would we want previous generations to have done with respect to the environment? Jacobs argues that we would want carrying capacity to be maintained over time so as to be able to enjoy an environment equal in quality to that of previous generations. And if we would want this for ourselves, we should ensure that this situation occurs for future generations.

Tietenberg (1994) suggests that the above relationship would emerge using the method set out by Rawls (1972) to establish the rules of justice, a method which places every person in an 'original position' behind a 'veil of ignorance'.

> In our context this approach would suggest a hypothetical meeting of all members of present and future generations to decide on rules for allocating resources among generations. Because these members are prevented by a veil of ignorance from knowing the generation to which they will belong, they will not be excessively conservationist (lest they turn out to be a member of an earlier generation) or excessively exploitative (lest they become a member of a later generation) . . .

> What kind of rule would emerge from such a meeting? Perhaps the most common answer is known as the sustainability criterion.

According to Tietenberg, the sustainability criterion requires that the well-being of future generations remains as high as that of all previous generations. This is to be achieved by ensuring that the value of the stock of natural capital should not decline over time. This would mean that one form of natural capital could be decreased only when another form of natural capital was increased as compensation.

More straightforwardly it can be argued that there is no reason to regard future generations as less important than ourselves, to regard them as anything other than morally equal. And given that this is the case, future generations should be entitled to the same quality of environment as that experienced by our current generation. This chapter adheres to the view that it is not acceptable to discount future lives, that future generations are morally equivalent to our own and that we therefore have a duty to maintain carrying capacity over time. We now turn to the question of how to measure carrying capacity.

The maintenance of carrying capacity

The environment acts as both a source of raw materials for economic processes and as a sink for the outputs of these processes. Maintaining carrying capacity means maintaining these source and sink capacities. Goodland, Daly and Sarafy (1993) set out 'operational principles' for maintaining sink and source capacities in the form of an output and input rule:

- *Output rule*
  Waste emissions into the environment must be within the assimilative capacity of that environment, i.e. its capacity to absorb waste without its future waste-absorptive capacity being diminished.
- *Input rule*
  (a) Renewables: harvest rates of renewable resources from an environment should be within the regenerative capacity of environment, i.e. its capacity to provide resources without its future capacity to do so being impaired.
  (b) Non-renewables: depletion rates of non-renewable resources should not exceed the rate at which renewable substitutes are developed.

The point to emphasise is that maintaining carrying capacity requires that goods and services produced to satisfy both our needs and desires will have to be produced from a finite quantity of resources – the quantity being limited by the operational principles set out above.

There can be no doubt that Europe is currently falling a long way short of meeting its obligation to maintain the carrying capacity of the environment. A recent report by the European Environment Agency (1995) on the state of the European environment (prepared at the request of the European Commission for its review of the 5th Environmental Action Programme) stated:

> The European Union is making progress in reducing certain pressures on the environment, though this is not enough to improve the general quality of the environment and even less to progress towards sustainability. Without accelerated policies, pressures on the environment will continue to exceed health standards and the often limited carrying capacity of the environment. Actions taken to date will not lead to full integration of environmental considerations into the economic sectors or to sustainable development.

So what quantity of resources are available to us? In 1995, a report prepared by the Wuppertal Institute for Friends of the Earth Europe (FoEE, 1995a) set out the first detailed attempt to quantify the reduction in resource use that would be necessary within Europe for it to play its part in maintaining global carrying capacity. This is referred to in the report as calculating Europe's 'environmental space'. Environmental space is defined as (FoEE, 1995b):

the global total amount of environmental resources: such as absorption capacity, energy, non-renewable resources, agricultural land and forests that humankind can use without impairing the access of future generations to the same amount.

Based on the assumption that each person has the right – although of course no obligation – to use the same amount of global environmental space, the report concludes that to live within their environmental space, European nations would on average have to cut their use of energy and non-renewable resources by approximately 80 to 90 per cent. Reduction figures have been calculated for the use of fossil fuels and nuclear power and for the use of various non-renewable resources (aluminium, cement, chlorine and pig iron). The figures calculated are not put forward as precise and definitive but aim to give a good estimation of the change required. Each national Friends of the Earth organisation in Europe has prepared or is now preparing a report calculating environmental space for their particular country using a similar approach to the report calculating environmental space for Europe as a whole.

It is noteworthy that the conclusions regarding environmental space set out above have been accepted by certain sections of the business community. The World Business Council for Sustainable Development (WBCSD) – an organisation formed in 1995 through the merger of the Business Council for Sustainable Development (BCSD) and the World Industry Council for the Environment (WICE)[1] – concedes that:

> industrialised world reductions in material throughput, energy use and environmental degradation of over 90 per cent will be required by 2040 to meet the needs of a growing world population fairly within the planet's ecological means.
>
> (WBCSD, 1993)

It has to be said that the severity of the reductions required are quite shocking when seen for the first time and are way beyond any reductions being suggested by any country within the EU or anywhere else in the world. According to the WBCSD this is hardly surprising as:

> only a small group of environmentalists, business executives, government officials and researchers have grasped the scale of change needed to achieve eco-efficiency. The business community is unclear about what is required of them.

This remark suggests that there is no doubt about the 'scale of change needed', whereas it is probably too early to be certain about the figures of 90 per cent or more being used by FoEE and WBCSD. However, this preliminary research should be taken as the starting point for more detailed work in this area. The calculation of environmental space is a vital step in our meeting our obligation to future generations, and government must therefore fund the research necessary to make this calculation and must initiate a detailed and open public discussion about the changes necessary to enable us to maintain carrying capacity.

2 THE NEED FOR GOVERNMENT INTERVENTION IN THE MARKET TO ENSURE THE MAINTENANCE OF CARRYING CAPACITY

Having established the obligation owed to future generations by this current generation, the means necessary for this obligation to be fulfilled will be discussed. Given that business is in the business of pursuing profit through the production of goods and services, this discussion begins by asking whether the pursuit of profit alone will enable business to bring about the reductions outlined above necessary to meet its obligations to future generations. In short, can 'the market' bring about sustainability?

Imagine a situation where there was no commitment to protecting the environment for future generations. In the absence of constraints on the use of the environment what would be the most profitable way to behave? There is an increasing body of case studies demonstrating how various businesses have been able to reduce their costs whilst at the same time improving their environmental performance (the famous 'win-win' scenario). Indeed Porter and van der Linde (1995) have argued that 'win-win' is the rule rather than the exception, that 'win-win' situations are pervasive throughout the economy. They argue that:

> pollution is a manifestation of economic waste and involves unnecessary, inefficient or incomplete utilisation of resources, or resources not used to generate their highest value.
>
> (Porter and van der Linde, 1995)

Hence improving resource efficiency is seen as a coherent strategy for reducing costs and gaining a competitive advantage, one which at the same time improves environmental performance. As Porter and van der Linde put it: 'At the level of resource efficiency, environmental improvement and competitiveness come together.'

So could 'win-win' take us to sustainability? Could the pursuit of profit through implementation of resource efficiency mean sufficient reductions in resource use to maintain carrying capacity? In order for this line of argument to be convincing we would have to show that:

(a) pervasive 'win-win' opportunities exist within the economy;
(b) although these 'win-win' opportunities have not so far been taken up, they will be from now on;

(c) 'win-win' opportunities will not disappear before the necessary reduction in resource use has been achieved; and

(d) increases in resource efficiency per unit of output are not outstripped by increases in output.

Arguments have been put forward contesting points (a) to (c) (see for instance Palmer, Oates and Portney, 1995) and these will be dealt with later. First of all, we will examine point (d). The crucial issue here is that whilst it may be most profitable to produce goods and services in the most resource efficient way possible, it will also be most profitable to sell as many of those goods and services as possible. Therefore there is every likelihood that growth in output will outstrip resource efficiency gains and that carrying capacity will therefore not be maintained.

Daly (1992) has argued that economics should be efficient, just and sustainable. Resources should be allocated efficiently within an economy, the goods produced should be distributed fairly and the scale of production should be sustainable. He goes on to argue that the market does only one thing: it solves the allocation problem. 'The policy instrument that brings about an efficient allocation is relative prices determined by supply and demand in competitive markets.'

However, whereas the market has a policy instrument governing allocation, it has no such instrument governing scale – the market has no built-in tendency to grow only up to the scale of aggregate resource use that is sustainable. Hence, scale must be determined 'by a social decision reflecting ecological limits' (Daly, 1992).

In order to meet our obligation it is therefore necessary to take a 'social decision' to limit the scale of economic activity to a size that does not degrade the environment over time. In order for such a decision to be made, a willingness has to be present amongst the various actors to meet this obligation. Three possible scenarios suggest themselves:

- Both business and government are committed to fulfilling their obligations and cooperate together to maintain carrying capacity.
- Government is committed and imposes its will on a recalcitrant business community.
- A committed business community drags with it a less committed government.

Obviously the first scenario is by far the most desirable and is the one which must be actively promoted. From the point of view of business, a commitment to meeting its obligations to future generations does not mean giving up on the idea of making a profit. However it does mean a recognition of the fact that making profit has to be constrained by this obligation. If making profit and protecting the environment come into conflict it is the pursuit of profit that must give way. This is the view advocated by the WBCSD (1993) which believes that:

30

The overarching eco-efficiency task for business is to fulfil the needs of all the world's people within the planet's means . . . The eco-efficiency bottom line is to make profits within the Earth's carrying capacity.

And it is optimistic about business prospects suggesting that:

the businesses which have prepared for these revolutionary changes will not only avoid extinction, but will thrive on the rapidly growing demand for eco-efficient products and services.

The WBCSD believes that the pursuit of eco-efficiency will:

provide a long term framework for unleashing the innovative powers of the corporate sector and opening up whole zones of profit making opportunities for new and old companies alike.

Self-regulation or government regulation?

It has been argued above that some form of market intervention is necessary to ensure that business plays its part in achieving the maintenance of environmental carrying capacity. This intervention could take two forms:

- the business community regulates itself, i.e. self-regulation;
- government intervention in the market.

Even if the business community was completely committed to preserving carrying capacity, each firm merely concentrating on its own environmental performance would not constitute effective self-regulation as this would not guarantee that the aggregate environmental impact of all firms would not harm carrying capacity. What is commonly thought of as self-regulation, i.e. 'compliance-plus' activities such as the implementation of environmental management systems (e.g. ISO 14001, EMAS) and signing up to various codes of conduct (e.g. the ICC Charter, the Ceres Principles) would therefore not be sufficient to ensure that our obligation to future generations is met.

In order for business to be effectively regulated, environmental space must be quantified and mechanisms put in place to ensure that environmental space is not breached. It seems quite unrealistic to expect business to perform this role. It is government which is the appropriate institution to coordinate the activities of the business community as a whole, i.e. to quantify environmental space and to establish the appropriate policy regime to ensure that environmental space is not breached.

3 EU POLICY: GROWTH, COMPETITIVENESS AND ENVIRONMENTAL SPACE

It has been argued in section 2 that government intervention in the market is necessary to ensure the transition to living within environmental space. Yet in section 1 it was shown that at present, EU policy is if anything taking us in the opposite direction. To reiterate the findings of the European Environment Agency (1995), although the EU has made progress in reducing certain pressures on the environment, 'this is not enough to improve the general quality of the environment and even less to progress towards sustainability'. Given that the current level of intervention has failed to get us anywhere near to living within environmental space, more stringent intervention is required. How much more stringent? Enough to ensure that we reduce our resource use to a level at which environmental carrying capacity is maintained (i.e. that the output and input rules set out previously are obeyed).

Whilst the EU has a duty to maintain carrying capacity, it is also advocating continuing economic growth. There has been much discussion in Europe recently about the need to boost European competitiveness and reduce the high levels of unemployment which currently exist. The White Paper on Growth, Competitiveness and Employment (EC, 1993) advocates a minimum annual growth rate in the EU of 3 per cent (coupled with an annual increase in employment intensity of between 1 and 1.5 per cent) in order to create 15 million new jobs and halve the present rate of unemployment by the year 2000.

Deregulation and reregulation

One of the measures that has been put forward by the EU as being able to contribute to meeting this growth target is legislative and administrative simplification. For instance the Molitor Report (EC, 1995), an expert report on legislative and administrative simplification prepared for the Commission, states that:

> Many well known factors influence the degree of competitiveness of European companies and therefore their capacity to create and to increase employment . . . Legislative simplification is therefore only one aspect amongst those which can increase competitiveness and employment . . . The elimination of unnecessary legal and adminis-trative burdens and simplification of regulatory frameworks is an important contribution in creating the conditions in which employ-ment goals can be realized and the global competitiveness of European business enhanced.

The aim of legislative and administrative simplification is, where appropriate, to revise existing regulations (reregulation) or to replace them with alternative mechanisms such as voluntary agreements, market mechanisms or self regulation (deregulation) in order to reduce the cost to business of meeting policy targets. It is hoped that this reduced cost will make goods and services cheaper and that this will result in an increase in demand for these goods and services. This increased demand will lead to firms within the EU taking on new employees, thereby reducing unemployment.

Is the 3 per cent rate of growth advocated by the EU compatible with its duty to ensure a transition to living within environmental space? In order to achieve economic growth (i.e an increase in the number of goods and services produced) whilst making the transition to living within environmental space (i.e. reducing the use of natural resources) goods and services produced within the economy must embody fewer natural resources (i.e. dematerialisation must occur). So what level of dematerialisation must occur for the 3 per cent growth rate to be achieved whilst reducing the use of non-renewable resources by 90 per cent (the figure suggested by FoEE as necessary to make the transition to living within environmental space – see section 1). If the EU economy were to grow at 3 per cent per year for 100 years whilst at the same time making the transition to living within environmental space (and remain so), then at the end of that 100-year period the economy would be almost 20 times as big as it is today and would need to dematerialise by a factor of approximately 200. (The relationship between growth and dematerialisation is examined in detail in the appendix.) The prospect of such an enormous dematerialisation being achieved does seem somewhat unlikely.

Given this current generation's duty to bring about a transition to living within environmental space, growth, if it is to occur, can do so only if it in no way disrupts this transition. If the dematerialisation mentioned above is not possible and the EU's desired rate of growth therefore proves to be incompatible with the transition to living within environmental space then the rate of growth must be constrained by the need to make this transition. (It is morally unacceptable for the reverse situation to occur, i.e. for a commitment to be made to achieving a particular rate of growth and for the transition to living within environmental space to be constrained by this growth commitment. The transition may not be possible at this particular rate of growth and so we would be unable to fulfil our moral obligation to future generations.) The above argument is summarized in Figure 2.1.

De/reregulation is a tool to enhance the efficiency with which policy goals are achieved, i.e. it reduces the cost to business and society of meeting a given policy goal. These efficiency gains explain the strong support within the EU for broadening the range of environmental policy instruments used to include not just traditional command-and-control regulation but also economic instruments, negotiated agreements and market-based instruments (e.g. environmental management systems and eco-labelling).

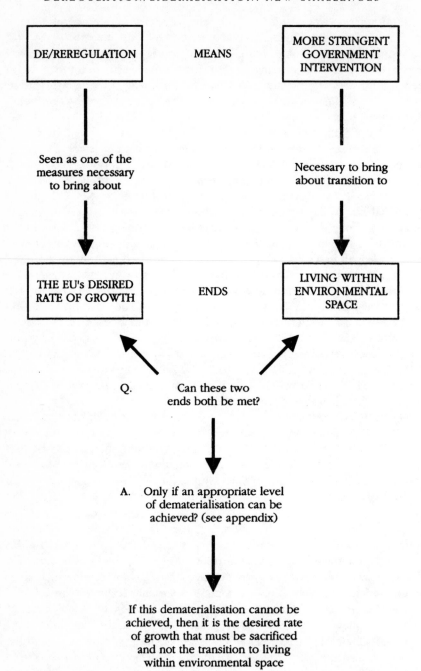

Figure 2.1 Is growth compatible with living within environmental space?

It has been argued here that in order to ensure that we move to living within environmental space, government intervention in the market is needed and given that the current level of intervention has failed to get us anywhere near to living within environmental space, more stringent intervention is required. Hence any reregulation and deregulation that occurs must take place within the context of sufficiently stringent government intervention. In other words environmental de/reregulation is a tool for ensuring an *efficient* transition to living within environmental space[2] and it is in this context rather than as a tool for promoting growth that deregulation should be seen.

Stricter regulation improves competitiveness?

What will be the effect on competitiveness of the more stringent government intervention mentioned above? Porter and van der Linde (1995) have argued that competitiveness can be brought about through the implementation of strict, 'properly designed' environmental regulation. They claim that although 'win-win' situations are pervasive throughout the economy, by and large firms will not discover them if left to their own devices and that market intervention of some kind is needed to enable firms to take up these opportunities. If such opportunities exist what stops firms from taking advantage of them? Porter and van der Linde (1995) suggest that firms have limited knowledge of the ever changing technical opportunities for cost saving and suffer from organisational inertia and control problems reflecting the difficulty of aligning individual, group and corporate incentives. Hence regulation is needed to enable uptake of these cost-saving opportunities. Properly designed regulations, it is argued, can trigger innovations that will fully or more than fully offset the costs of complying with them. Regulation is viewed in a positive light, as the 'innovation offsets' which arise as a result can lead to a lowering of cost and an advantage over firms in other countries not subject to such regulations. As the above authors put it 'strict environmental regulation can be fully consistent with competitiveness'.

Palmer, Oates and Portney (1995), on the other hand, maintain that whilst 'win-win' situations do on occasion exist, they are not pervasive throughout the economy. They reject the view held by Porter and van der Linde that the private sector is one which systematically overlooks profitable opportunities for innovation and which needs the intervention of regulatory authorities to provide the needed incentives for cost saving and quality improving innovations that competition apparently fails to provide. By and large, they argue, most opportunities for cost-saving environmental improvement are taken up – firms tend to pick up most $10 bills that they find lying around.

Given that this is the case, there are, they argue, not a huge number of remaining cost-cutting opportunities for environmental regulation to stimulate. Therefore, far from regulation being a cost-cutting exercise as

Porter and van der Linde maintain, it is in fact something that entails costs for most firms. With literally hundreds of thousands of firms subject to environmental regulation it would be hard not to find instances where regulation has seemingly worked to the polluter's advantage. Given that all firms are not perched ever-vigilantly on their efficiency frontiers there will always be instances where regulation leads to the discovery of cost-saving innovation. However, compiling cases where this has happened in no way establishes a general presumption in favour of this outcome. It would be an easy matter, claim the authors, to 'assemble a matching list where firms have found their costs increased and profits reduced as a result of (even enlightened) environmental regulations, not to mention cases where regulation has pushed firms over the brink into bankruptcy'.

Rather than relying on case studies, the authors seek to demonstrate their case that regulation means increased costs for firms (due to the absence of pervasive 'win-win') by looking at statistics collected by the US government. Each year the Environmental Economics Division of the Commerce Department's Bureau of Economic Affairs (BEA) compiles data on pollution abatement and control expenditures in the United States. In 1992, according to BEA pollution abatement and control expenditures came to $102 billion. Meanwhile, the Environmental Protection Agency, whose methods of calculating data in this area are slightly different, estimates the total cost at $135 billion.

In addition to estimates of environmental spending, BEA also estimate the magnitude of the 'offsets' using data from the Bureau of Census survey of manufacturers. For 1992, BEA estimated that cost offsets for the US amounted to $1.7 billion, less than 2 per cent of estimated expenditures. One possible criticism of these estimates of offsets is that certain kinds of offsets in response to more stringent environmental regulation are not easily reportable and hence do not find their way into the BEA estimates. However, even if the estimate offsets were tripled or quadrupled there is still a yawning chasm between offsets and compliance costs. The above argument is supported by Cairncross (various, 1994) who argues that:

> Most of the $10 bills to be had by reducing pollution or saving energy have either been picked up or can be achieved only at a cost . . . It is not surprising that tougher standards impose costs on companies. The aim of such standards, after all, is to force polluters to internalise costs previously inflicted on society. Or future generations inherit them. Environmental policies that are worth pursuing should be introduced for their own sake. To try to improve competitiveness by raising environmental standards is to risk the fate that typically awaits those who try to ride two horses at once.

Whilst it is of interest to establish whether or not increased environmental regulation leads to increased competitive advantage, the answer to this question should have no bearing on the pursuit of environmental protection. We are under an obligation to maintain carrying capacity for future generations and this must occur whatever the consequences for business competitiveness – as Cairncross argues, environmental protection must be pursued for its own sake. If it is the case that strict, properly designed regulation does in fact reduce costs and lead to a competitive advantage in the country/region where it is applied, all well and good. If it does not and costs to business increase, then it must be accepted that this is the price that must be paid in meeting our obligation to preserve environmental carrying capacity.

CONCLUSIONS

This chapter has argued that considerations of social justice require our current generation to maintain environmental carrying capacity for future generations. Having established the maintenance of carrying capacity as a policy goal, it is necessary to calculate the changes in resource use necessary to reach this goal. Preliminary research by Friends of the Earth Europe has indicated that in order to maintain carrying capacity (or, to use their terminology, to live within environmental space), it may be necessary to reduce consumption of fossil fuels and non-renewable resources within the EU by 80 to 90 per cent. Performing as accurate a calculation of environmental space as possible is an integral part of our meeting our obligations to future generations and government is the most appropriate agency to carry out this task. Hence it is crucial that government devotes the necessary resources to this task.

Having calculated the change in resource use required to live within our environmental space, it is necessary that the appropriate mechanisms be put in place to ensure that these changes occur. It has been argued here that the market has no inbuilt mechanism for ensuring that economies operate at a sustainable scale and that some form of intervention in the market is necessary to bring this about. The compliance-plus activities of business commonly regarded as constituting self-regulation, whilst worthwhile, are not sufficient to guarantee that environmental carrying capacity is maintained. For self-regulation to be successful in maintaining carrying capacity, business would have to calculate environmental space and design and implement measures that enable it to play its part in ensuring that this space was not breached – not a realistic proposition. Government, not business, is the appropriate institution for performing these duties.

Given that this is the case, what policy measures are open to government? It has been argued in this chapter that we are at present a long way from reaching sustainable development and that current EU policy is not taking us closer to this goal. Hence more stringent intervention is needed by government in order to make the transition to a sustainable economy.

Obviously, the most suitable forms of intervention should be used by government and hence the case for reregulation and deregulation should be examined. It has been argued that strict environmental legislation may in fact bring about lower costs for business. If this is the case all well and good. If not, then the higher costs entailed must be accepted as the price to be paid in meeting our obligations to future generations.

ACKNOWLEDGEMENTS

The author would like to thank Martin Enevoldsen for his helpful suggestions regarding the first part of this chapter and Andy Gouldson for his encouraging comments. Thanks also to Steve Dunbar for his most helpful technical support during the final stages of preparing this chapter. Special thanks to David Fleming for the time and effort spent commenting on my draft manuscript. As ever, his comments were perceptive, helpful and gratefully received. And finally thanks to Ute Collier for her patience during the editing of this chapter.

NOTES

1 WBCSD has a membership of 124 companies. The membership consists of companies from all five continents. According to its mission statement WBCSD aims to be 'a leading business advocate for issues regarding the environment and sustainable development' and to 'participate in policy development to create the right framework conditions for business to make an effective contribution towards sustainable development' (Wyburd, 1996). See this reference for further information on WBCSD.
2 Whilst deregulation is concerned with efficiency, making the transition to living within environmental space is an issue of scale. The importance of efficiency and scale and the relationship between the two is discussed in Daly, 1992.

APPENDIX

What would need to happen in order to combine the White Paper's aspirations for growth with our duty to move to living within environmental space? The answer to this question is discussed in this appendix.

Environmental impact is frequently represented by the equation $I = P \times A \times T$ where

I = environmental impact;
P = population;
A = affluence (output consumed per capita); and
T = technology (environmental impact per unit of output).

The Wuppertal Institute (FoEE, 1995a) point out that the products we use provide us with services and it is these services that contribute to our well-being. Hence the Institute defines affluence in the above equation as services consumed per person (S/P). As the Institute uses material input to the economic system as a proxy for environmental impact, technology is defined as material input per service unit – referred to as MIPS. The above equation can therefore be restated as

$I = P \times S/P \times MIPS$

MIPS can be reduced by producing goods that embody less resources and/or are more energy efficient. Making goods that last longer and that are repairable also reduces MIPS as more services are derived from a product. MIPS can also be reduced by more people using a product (e.g. car pool schemes) or by providing the same service in a less resource intensive way (e.g. public as opposed to private transport). Hence MIPS can be reduced through improvements in resource efficiency and product design and by behavioural changes. The following calculations show the reduction in MIPS necessary for a transition to living within environmental space at various rates of growth.

Let us assume there is zero economic growth, i.e. that population (P) × services consumed per person (S/P) is constant. What reduction in MIPS is needed to make the transition to living within environmental space? If we assume that this transition entails a reduction of 90 per cent in the use of non-renewable resources (I), i.e. the figure calculated for Friends of the Earth Europe by the Wuppertal Institute, then given that P × S/P remains constant, MIPS has to be reduced by 90 per cent, i.e. a dematerialisation of a factor of 10 (see Table 2.1). Preliminary research by the Wuppertal Institute suggests that the technology exists to bring about this degree of dematerialisation (Tischner and Schmidt-Bleek, 1993).

However, much bigger reductions in MIPS are required if we are to make the transition to living within environmental space and sustain a growth rate of 3 per cent (the acceptable EU minimum). If we assume the transition to environmental space is to take 50 years, then at an annual growth rate of 3 per

Table 2.1 Required rates of dematerialisation

Rate of economic growth (%)	No. of years	Resulting services (current level of services = 1)	Required MIPS (current MIPS = 1)	% dematerialisation	Dematerialisation factor
0	50	1.00	0.100	90.0	10.0
1	50	1.65	0.060	94.0	16.5
2	50	2.69	0.037	96.3	26.9
3	50	4.38	0.022	97.8	43.8
3	100	19.21	0.0052	99.5	192.1

Source: Adapted from Friends of the Earth Europe (1995a)

Is this degree of MIPS reduction feasible? Can every service we enjoy be produced with 192.1 times less non-renewable resources than today? Although one cannot say for certain, it does seem somewhat unlikely. If 3 per cent growth within the EU is desired then the approach has to be to attempt to produce the amount required whilst making the transition to living within environmental space. If this can be achieved, all well and good. If not then the particular growth rate required must be forsaken and it must be accepted that it is just not possible to obtain such a rate of growth whilst fulfilling our moral obligations to future generations.

cent the GDP would be almost 4.5 times larger at the end of the transition period than at the beginning (4.5 times as many services). Hence MIPS has to be reduced by a massive 97.8 per cent or a factor of 43.8! And if growth continued for a further 50 years, the GDP would be 19.21 times as big as at the beginning of the transition period. In this case a dematerialisation of a mind-boggling 99.5 per cent, i.e. by a factor of 192.1 is required!

REFERENCES

Adams, J. (1995) *Cost–Benefit Analysis: Part of the Problem, Not the Solution*, Oxford: Green College Centre for Environmental Policy and Understanding.

Costanza, R. and Daly, H. (1992) 'Natural capital and sustainable development', *Conservation Biology*, 6(1), pp. 36–46.

Daly, H. (1992) 'Allocation, distribution and scale: towards an economics that is efficient, just and sustainable', *Ecological Economics*, 6, pp. 185–193.

EC (1993) *Growth, Competitiveness, Employment – The Challenges and Ways Forward into the 21st Century* (White Paper), Supplement 6/93, Bulletin of the European Communities.

EC (1995) 'Report of the Group of Independent Experts of Legislative and Administrative Simplification', *COM* (95) 288 final/2.

European Environment Agency (1995) *Environment in the European Union 1995*, Copenhagen: EEA.

Friends of the Earth Europe (1995a) *Towards Sustainable Europe – The Study*, Brussels: Friends of the Earth Europe.

Friends of the Earth Europe (1995b) *Towards Sustainable Europe: A Summary*, Brussels: Friends of the Earth Europe.

Goodland, R., Daly, H. and Serafy, S. (1993) 'The urgent need for a rapid transition to global environmental sustainability', *Environmental Conservation*, 20(4), pp. 297–330.

Jacobs, M. (1991) *The Green Economy*, London: Pluto Press.

Layard, R. and Glaister, S. (1994) *Cost–Benefit Analysis*, Cambridge: Cambridge University Press.

O'Neill, J. (1993) *Ecology, Policy and Politics*, London: Routledge.

Palmer, K., Oates, W. and Portney, P. (1995) 'Tightening environmental standards: the benefit-cost or the no-cost paradigm?', *Journal of Economic Perspectives*, 9(4), pp. 119–132.

Porter, M. and van der Linde, C. (1995) 'Towards a new conception of the environment–competitiveness relationship', *Journal of Economic Perspectives*, 9(4), pp. 97–118.

Rawls, J. (1972) *A Theory of Justice*, Oxford: Oxford University Press.

Tietenberg, T. (1994) *Environmental Economics and Policy*, New York: Harper Collins College Publishers.

Tischner, U. and Schmidt-Bleek, F. (1993) 'Designing goods with MIPS', *Fresenius Environmental Bulletin*, 2(8), pp. 479–484.

Various authors (1994) 'The challenge of going green', *Harvard Business Review*, July–August 1994.

World Business Council for Sustainable Development (1993) *Getting Eco-efficient*, Geneva: WBCSD.

Wyburd, G. (1996) 'BCSD + WICE = WBCSD', *Business Strategy and the Environment*, 5(1), pp. 48–50.

3

DEREGULATING ENVIRONMENTAL LAW IN A PERSPECTIVE OF STIMULATING KNOWLEDGE GENERATION

Karl-Heinz Ladeur

PRELIMINARY REMARKS ON THE CONCEPT OF REGULATION

Speaking of the merits and flaws of deregulation clearly presupposes a consideration of regulation – a concept to which deregulation remains linked, although antithetically. Selznick (1985), an American legal theorist, provides a helpful description in his definition of regulation as a sustained and focused control exercised by a public agency over activities that are socially valued.[1]

An environmental policy perspective requires, in addition to this formulation, that we distinguish basically between two strands of regulation from the outset. The first is intended to compensate for a lack of market control (especially in the domains of so-called natural monopolies, such as the large communications and electricity distribution networks, to name but two). The second, apparently more central to our topic, aims at safety control through technical rules imposed on activities (production of goods, etc.) which in other respects (price, quality, etc.) are controlled by market forces (Noll, 1982; Grindley, 1995). The first strand of regulation, which in many ways has found in Europe a different but functionally equivalent solution in 'public services' (especially the French concept of '*service public*'), is still prominent in this domain (Boiteux, 1996; Collier, 1995).

This first strand has come under the pressure of growing technological flexibility, with the universal service guaranteed by a single integrated network being challenged by differentiated and multiple networks (Henning and Klodt, 1995). It is beyond the scope of this paper to go into detail here on the problems related to strategies for unbundling, for example of electricity networks, currently being implemented by the European Union.[2] However, we will see that interrelationships exist between this domain of regulation and the safety base alternative. This brings us to the role played by public authorities in exercising a 'sustained and focused control' over economic activities, as described by Selznick. This component of the concept of regulation makes it clear that we are not confronted with just a more technical, specialised form of rule-making, but that regulation presupposes 'moving targets', an evolving knowledge base and acquaintance with specific problems concerning the technological and economic conditions of the regulated domain. This interdependence between private and public decision-making explains why the growing flexibility of technology inevitably has some strong repercussions on safety regulation as well.

This assumption may help us gain a clearer idea of the relationship between technical knowledge, legal frameworks, private production and public control in the past. In this way, we can establish a frame of reference which might on the one hand enable us to form a clearer impression of the ongoing transformation within safety control. On the other hand, it can help us understand the simple alternative of either defending the public regulatory logic or doing away with its clumsiness.

REGULATION AND ITS FRAME OF REFERENCE: EXPERIENCE-BASED AND RULE-ORIENTED

In the past, regulation was characterised by an assumed stable framing of the knowledge base, which allowed for a balanced coordination between private action and public control (Ladeur, 1995). The core component of this knowledge base was *experience*, conceived of as a set of practical rules formulated on the basis of *single events*, particularly accidents (Rasmussen, 1995), that are empirically observable (with the exception of those very basic natural laws and regularities of natural science considered to be exempt from change). It was held that experience as a form of knowledge comprised a mode of self-transformation, which was at the same time channelled and limited by the generative rule of admitting only empirical information and by the assumption that the continuous development of events would lead to a continuous evolution of experience. This means that it would exclude the emergence of disruptive phenomena which devalue the evolution of rules themselves.

This assumption allowed for a separation of work and production rules on the one hand, and safety control rules on the other (Rasmussen, 1995). These safety rules could be imposed in a hierarchical manner within organisations. Working with high-pressure containers or with inflammable substances was supposed to be steered by general safety standards formulated on the basis of empirical observation of accidents. Public safety control was based on the concept of prevention of danger, a concept which reflects the intertwining of continuity and change that is characteristic of experience as a set of rules for practical action. It presupposed a stable image which separates the normality and integrity of public and private goods from harm, thereby serving as indicators of the necessity for public action. But at the same time, the concept of danger is linked to its antipode; deviation from the normal code of conduct. This notion ties in with the meaning of experience: it does not exclude the possibility of harm caused by actions, but imposes rules of conduct in taking risks. This approach therefore assumed that accidents happen, even despite obedience to rules, but that we need a stop-rule limiting control efforts and allowing for experimentation with new technical designs, the introduction of new components into established systems, and so on. This stop-rule, linked to piecemeal engineering, was considered to be viable because the distributed ongoing development of technology would limit the consequences of accidents and at the same time allow for learning (Ladeur, 1995).

The different components of the safety control system linked quite elegantly together. Technology develops continuously, drawing on experience as a knowledge base acceptable to all, new experimental actions are distributed over a decentralised technical system, and safety control is codified in a number of separate rules imposed on work processes (Rasmussen, 1995). This is the private side of the regulatory field; public control can easily be linked to experience-based rules of conduct, which can be used to make the concept of danger specific. The causal and action-based rationality of this conception can readily be seen, for instance, in the problem of a general degradation of air quality in industrialised regions at the end of last century. When there were hundreds of polluters, each contributing in a minor way to pollution, no causal link could be established between single action and identifiable and separable harm which could be attributed to a specific firm. This causal construction ties in with the normative assumption that in this case no line can be drawn separating the domain of the normal from what must be regarded as a deviation.

One can thus draw the preliminary conclusion that this traditional or even classical model of safety regulation is rather more sophisticated than is implied by reference to the commonly used name of 'command-and-control' approach. But what one should keep in mind is that it is essentially a linear equilibrium model. Innovation and change, empirical and normative evolutionary steps, appear as fluctuations which do not call into question the preservation of a point of stability that is repeatedly re-established through a process of self-stabilisation of empirical regularity and legal normativity.

THE EVOLUTION OF REGULATION AND THE GROWING COMPLEXITY OF ITS KNOWLEDGE BASE

The rather simplified model I have described can serve as a starting-point in helping us understand the further evolution of the regulatory process. In my view, this evolution can be described as an example of a kind of institutional muddling-through, which consists in a strategy of reformulating existing, as well as introducing new, components into a system without consideration of the sustainability of the overall model. This is something lawyers are quite familiar with; it is completely legitimate to experiment with the viability of a system and to attempt to check its adaptability to changing environments. And this is, in my opinion, precisely what is happening in the ongoing regulatory process. To give but a few examples: the concept of danger has been more and more specifically formulated and normatively differentiated, especially in environmental legislation. It has even been supplemented by the new concept of precaution (Jans, 1995).[3] Furthermore, the range of public and private goods used as indicators of harm has been broadened to include 'goods', such as climate protection and the sustainability of ecosystems, which do not fit into the traditional scheme of separable goods attributable to individuals (property) or the state.

On the other hand, the knowledge base of public decision-making has also undergone increasing specification and differentiation; the dynamic of knowledge generation, as well as the heterogeneity of different levels and paths of technological development must be taken into consideration. Regulators distinguish between a basic 'state of the art' and 'best available' technology or even the 'state of scientific knowledge' (Ladeur, 1997). The latter formulae requires a dynamic evolution that is not regarded as part of a spontaneous development of a technology, but rather as something which may be prone to strategic intervention, either public or private. Reference to the 'state of scientific knowledge' (which is not yet put into technological practice) assumes a much more direct link between technology and research than was the case in the past. In former times, research also contributed to reproducing and deepening a common knowledge base accessible to all and not directly linked to technical practice. It was held to be part of a public cultural and educational system, which also served as a basis for the qualification of engineers, who would then try to apply it in practice. Research was intended to contribute to the understanding of universal rules which were regarded as being separate from practice (Wagner, 1995).

The complexity of this differentiated and heterogeneous knowledge base is reflected in the rise of numerous procedures and committees (Joerges, 1996) established for the formulation of rules that are increasingly abstracted from private action itself. Standards are set by specific private bodies or administrative committees and the role of evaluation and fact-finding becomes much more problematic (Otway and von Winterfeldt, 1992). This development can be systematised as a tendency to organise the process of knowledge generation, either for the internal self-control of correction processes or for the sake of public decision-making. This process has been adapted to a certain extent to many production processes in big firms, but it cannot really come to grips with the fact that technology has long-range effects (such as nuclear energy) or is difficult to manage due to insufficient knowledge (biotechnology). It is not possible to explore this area in detail here, but what appears to be crucial is the hypothesis that the system of coordination between the components of classical regulation has been overburdened by this development.

CURRENT POLITICAL AND LEGAL APPROACHES TO REDUCING REGULATORY COMPLEXITY: THE GERMAN CASE

Differentiation of the regulatory system has become so complex that a new comprehensive remodelling of the entire network of interrelationships between technology and internal safety control, the knowledge base and normative framework, and private organisational and public administrative processes has become necessary.

This is the core element of rationality evident in current attempts to introduce greater flexibility into administrative procedures. But a rather short-sighted approach has dominated discussions to date within the legislature. In Germany, for example, the government wants to put an end to lengthy and time-consuming licensing procedures in order to strengthen the competitiveness of industry. Parliament has rather haphazardly amended many environmental laws and thus accelerated the law-making process rather than administrative decision-making procedures. In particular, the Pollution Control Act was amended in 1994, again in 1995, and a new amendment has been proposed in 1996.[4]

The problem with these amendments lies in the fact that they seek to improve a clumsy procedure in order to speed up the licensing process, without really dealing with its complexity. The end-result, therefore, is that provisional and broad 'framework licences' are delivered, which merely shifts the problem to the future. Firms can get to work earlier, but at the same time are forced to sacrifice the stability of licensing decisions in their quality as administrative acts. Whether this is really acceptable for industry itself remains rather dubious. This approach amounts to a stop-go policy, which will create considerable disruptive effects in the legal system and will particularly undermine coordination between different principles of regulation.

In my view, this development highlights how important it is to elaborate an improved description of the disaggregating tendencies that the classical regulatory model is exposed to.

A particular suggestion by a German Expert Commission has been to introduce an optional procedure replacing the duty to apply for an administrative licence for an industrial installation by a simple duty to inform the authorities of the essential characteristics, e.g. of a new, potentially harmful, chemical plant. The administration would then have the discretion to insist on a formal procedure. One of the advantages of this type of decision-making would consist of creating competition between public agencies.

The 'framework licence' is another alternative within a broad range of different possibilities of combining public and private involvement in the control of environmental damage caused by industry. It would not eliminate all public control but distinguish between different steps and types of control and allow for reservations to impose restrictions at a later stage. A further potential element of flexibilisation would consist of the introduction of an option to allow for the qualified, expertise-based, private self-evaluation of risks as a basis for licensing procedures.

A SECOND LOOK AT THE TRANSFORMATION OF TECHNOLOGICAL PROCESSES

The reason why the command-and-control approach cannot be regarded as the only regulatory model can be found in the deep transformation that technological processes have undergone in recent years. They no longer follow stable paths within established paradigms of, for example, chemical process engineering. The design of a new chemical plant is no longer linked to experience-based knowledge and established trajectories. It does not draw on present practical knowledge but is based on research and theoretical knowledge, with the design to be tested and revised in the construction and operation process itself. This is also the case for biotechnology, and in general for production processes. These areas are characterised by a close link between theoretical design and its practical 'application', which at the same time is structured as a knowledge-generation process linked in a circular way to the design itself.

This element of systematic knowledge generation through a technology which is based on research, and simultaneously draws on feedback to practise, undermines the clear separability of rule-based knowledge and its continuous application in practice. Furthermore, rapidly changing production methods and innovative, 'intelligent', knowledge-generating enterprise strategies (Nonaka, 1991; Nelson, 1991) are scarcely accessible to administrative control, at least not within its classical frame of reference. This means that the knowledge process itself is less and less accessible to external supervision according to rules in the strict sense. It is for this reason that many authors, especially economists, are in favour of a systematic change of approach to, say, pollution control (Rajah and Smith, 1994; Rehbinder, 1995; Irwin and Vergragt, 1989). They recommend that we leave aside all the technical problems of regulation, meaning regulation of behaviour, and shift to the alternative of setting outcome-oriented targets, with the practical problems of action to be taken left to firms.

These goals should either be attained through economic instruments such as ecotaxes or the establishment of a market for pollution rights, or strict liability (Mitts and Holzman, 1994). These alternatives are of course not to be bluntly rejected. However, the knowledge problem cannot be coped with in a simple, uniform manner. Moreover, we must bear in mind that the high level of flexibility attained in industry is not restricted to technology alone; it covers organisational structures as well. Risky productions can easily be outsourced and transferred to under-capitalised firms. And still the collection of knowledge as a public good, especially the collection, recombination and redistribution of risk-knowledge spread over a multiplicity of enterprises, must be managed by public authorities once the spontaneous character of public-learning processes based on experience is interrupted.

FROM INSTRUMENTAL TO COGNITIVE RATIONALITY

Other similar approaches try to supplement efficiency-oriented instrumental rationality with a 'cognitive' one (Munier, 1995), and take into consideration the productivity of networks of decisions and the patterns they generate. The sequential combination of decisions, which are being adapted to processes of experimentation with constraints and new options, must also be accessible to legal control designs. This is possible only by introducing procedural rules for self-observation of enterprise organisation, as their management cannot be accessed by spontaneous trial-and-error processes alone. This is because the development of inter- and intra-organisational long chains of actions, and complex networks of relationships, undermines the transparency of markets and their monitoring capacity. The element of monitoring must be established and stabilised through procedural forms because the public should take an interest in conserving the innovation potential of the economy, including evaluation of the risks presented by new institutions.

At this point we are again confronted with the dissolution of the dividing line between private and public. This is due to the fact that competition can no longer be assumed to fulfil its functions once its frame of reference, established by presupposing an equilibrium model, has become ambivalent. The collapse of an enterprise can no longer start from the logic of an 'economy of scale', which assumes that the achievements of a disappearing enterprise will be taken over by a better, i.e. bigger, enterprise. In an 'economy of scope' evolving through the flexible recombination of specialised forms of knowledge, the collapse of an enterprise can be equivalent to a loss of knowledge that cannot easily be compensated (Eberl, 1996). The warning capacity of the financial market should not be overestimated either. This does not mean that inefficient enterprises should be conserved through subsidies, and so on. It is more a question of establishing a new description of the economy and its functioning for the purposes of public regulatory decision-making. The lack of flexibility of state regulation is no argument in favour of a position claiming that the functions fulfilled by the state in the past can now be easily taken over by industry itself. Industry too has to cope with uncertainty.

Drawing on economic notions of cognitive rationality, a broader self-description of the functioning of enterprises could be linked to a reformulation of the problem of flexibility in legal decision-making. The new self-reflective elements of economic decision-making could tie in with a broader, more flexible cooperative approach in legal decision-making, open to learning processes and facilitating learning processes in the economy.

The law has to observe and revise its own frame of reference within a new cognitive rationality focusing on proceduralisation. The legal process must systematically take into account its incompleteness and be open to processes of discovery of, and experimentation with, new options for private self-organisation and their evaluation.[5] Enterprises and markets can no longer guarantee monitoring and management of complex knowledge in the established forms. New modes of observation[6] must accordingly be established to stimulate innovation and conserve the flexibility of inter- and intra-organisational networks of interrelationships, linked with the whole process of societal knowledge generation. Its reproduction is no longer guaranteed through relatively stable classical technological paradigms and the assumption of continuous enrichment of experience as a generally accessible body of knowledge. Moreover, this development cannot be steered by a state considered to be the centre of society. A functional equivalent to the classical reference to stable limits has to be sought, in order to develop a new paradigm able to take over the functions of the traditional model and to adapt to more complex forms of self-organisation of society.[7]

IN SEARCH OF A FUNCTIONAL EQUIVALENT TO THE CLASSICAL LIBERAL INTERRELATIONSHIP BETWEEN EXPERIENCE AND PRIVATE OR PUBLIC ACTION

My conclusion would thus be to propose a closer look at the transformation of technology and economic organisational processes, and to seek to establish a functional equivalent to the classical interrelationship between private action, knowledge and public safety control. We saw that the classical structuring paradigm consisted of *rules* as an intermediary component bridging the gap between private and public spheres. This can no longer work, but it should be possible to find a new approach which does not just passively adapt public decision-making to the constraints

49

produced by flexible economic processes. Rather, this new approach should test the possibility of redesigning state function of observation and control oriented towards public interests by reformulating the control task and shifting it to a higher level of abstraction that makes it no longer rule-based but *proceduralised*. What does this mean? The new target of control should not be rules separated from productive actions themselves. Instead, it should be a new concept which seeks to draw on the productive side-effects of the new paradigms of technology and economic organisation, in the same way that the classical approach did with experience-based technology and production processes and a market stabilised by a limited range of preferences. Flexibility of technology is a problem only for a rule-based approach, because of rapid change.

However, intelligent technology is also more open to the integration of multiple purposes, including safety, in a form of risk management (Rasmussen, 1995). This is rather different from the traditional separation between technological purposes and 'end of the pipe' control technologies. Technological development takes place nowadays in an 'inter-informational environment' of a learning organisation able to fulfil ever more complex and self-transforming tasks (see Hakansson, 1986). New forms, not of deregulation, but of reduced and simplified adaptive regulation, should foster a system design of technological processes which includes the problems of safety control and monitoring under the aspect of risk information and self-revision (Rasmussen, 1995; Ladeur, 1993). Firms should be obliged to develop a flexible risk observation and evaluation design themselves, including self-reporting and self-monitoring systems adapted to the management of multiple information. One could, of course, counter that this would be highly complex, and difficult for the administration to understand. This is indeed the case, but this approach would aim at a level of abstraction where not only firms, but administrations too, could learn to develop and understand methods and procedures of risk management, possibly with the help of private consulting firms.

On this basis, a new coordination between private and public could be established, which could be regarded as a functional equivalent to the classical paradigm of experience and its continual development. The idea consists in replacing rule-based regulation with a second-order form of procedural regulation in the sense of adopting a comprehensive reciprocal learning strategy as the basis for a new coordination between industry and administrative decision-making. This implies the assumption that a new risk-based design of complex production processes should be set up, drawing not only on the learning capacity of industry and some kind of self-enlightenment, but also on a design of production processes. These would systematically incorporate the problem of coping with risks and uncertainty, no longer trying to lay claim to be safety based on a single evaluation made before the production process is put into action. But risk management should be encouraged to lay open uncertainty and to introduce procedures of self-observation of the viability of assumptions made and necessary revisions to be incorporated on the basis of new information generated by practice.

Complex technological processes should have a built-in capacity for dynamic self-description, which could be more easily interpreted with the help of computer programmes able to handle the huge amount of potential information generated by complex production processes. This would amount to an idealised version of a production process evaluated not by public authority on the basis of stable rules linked to experience, but rather on the basis of its procedural sophistication in generating additional information on risks through its own 'application'. And this procedural element of self-observation and self-design could be the missing link which allows for a reciprocal relationship between private and public action. This conception is not based on confidence in professional risk management by industry, but rather on the possibility that private and public reference to the viability of methods and procedures of generation and management of risk information could lead to the setting up of a common, shared, second-order knowledge base. This would no longer be rule-based but procedure-oriented. In this respect, a functional equivalent to the relationship between rule and experience could consist in a procedural reflexive integration of observation and evaluation processes into flexible technological production processes.

To draw a parallel again with the classical experience-based paradigm, I would describe the new regulatory model as a non-linear disequilibrium model based on the assumption that the old separations have broken down and that regulation has lost its stable frame of reference. There is no way back to a simple constructive model, the conditions for which have vanished in the evolution of technology. But on the other hand, the complexity of technological processes cannot be reduced to their disruptive consequences for the regulatory system alone. They create new forms of procedural learning that the law can draw on in order to establish a new regime of the 'regulation of self-regulation', adapted to rapidly changing processes of self-organisation[8] which can no longer be dissolved into separable sets of individual behaviour to be judged according to experience-based rules. Instead, the whole system of interrelationships processing knowledge rather than material resources must be redesigned in a multifunctional way, integrating the generation of risk information.

This assumption does not exclude the continuing importance of the classical experience-based approach to regulation, nor does it rule out the possibility that in some domains it might be replaced by economic instruments. But one of the assumptions on which this chapter is based is the claim that the use of the latter can only be conceived of if the range of alternatives is limited, and the adoption of one of them, including its timing, can be left to the discretion of economic actors. But the dynamics of technological development can make any kind of technology-forcing, even by economic instruments, highly risky. It may create adverse effects unless it is borne in mind that the information which industry has available is more or less limited.

A misplaced incentive may lead to the adoption of a well-known technology based on a short-sighted trade-off, thus suffocating the development of much more promising and even less expensive process technologies in a longer time-frame. This is also an argument in favour of allowing for flexibility within licensing procedures (Ashford *et al.*, 1985), especially in the form of 'innovation waivers', despite the risk of failure. Under these conditions, the above-mentioned new idea of a 'framework' or 'global licence' discussed in Germany and implemented in the Netherlands could also be a very valuable innovative instrument. It should then be limited to complex innovative processes put into practice within a flexible proceduralised design. This should be linked to an intelligent component of risk observation and knowledge generation, able to legitimate the extension of guarantees of stability based on the evaluation of procedures, in the same way as the traditional approach draws on the reliability of a rule-based administrative act limiting revocation on the basis of new information given to the administration.

It should be clarified that the approach sketched out here would integrate the classical rule-based regulatory mode as a kind of first-order learning based on experience, whereas complex technological processes must be shaped by a procedure-based regulatory approach allowing for more complex processes of second-order learning related to mechanisms of knowledge-generation systematically linked to production processes. This type of regulation should not only stimulate creative and innovative technology but at the same time be open to self-revision, drawing on observation of a coevolutionary dynamic able to model new possibilities.

NOTES

1 See also Majone, G. (1994) and Majone, G. (1990).
2 See Boiteux (1996), Collier (1995).
3 Secretary of State for Trade and Industry exparte Durbridge, United Kingdom Queen's Bench Division, *Journal of Environmental Law* 7 (1995), p. 225; *Case Analysis* by David Hughes, pp. 238ff.
4 Compare project of law of 19 January 1996, article 1 No.2, §6, Pollution Control Act; general measures intended to speed up procedures now introduced into the (General) Law on Administrative Procedure of 10 September 1996 (§ 71aff.); see Koch (1996).
5 For self-organisational effects in the new *lex mercatoria* see Teubner, G. (1995).
6 For a theory of observation see Luhmann (1988).
7 For a theory of self-organisation in general see Allen (1988); with reference to economy see Arthur (1990).
8 For an interesting and innovative approach in the Netherlands see Frans van der Woerd in this volume.

REFERENCES

Allen, P.M. (1988) 'Dynamic models of evolving systems', *Systems Dynamics Review* 14 (2), pp. 109ff.

Arthur, W.B. (1990) 'Positive feedback in the economy', *Scientific American* (2), pp. 94ff.

Ashford, N.A. *et al.* (1985) 'Using regulation to change the market for innovation', *Harvard International Law Journal* 26 (1), pp. 419ff.

Boiteux, M. (1996) 'Monopole ou concurrence dans l'électricité?', *Le Monde*, 3 May, p. 14.

Collier, U. (1995) *Electricity, Privatisation and Environmental Policy in the UK: Some Lessons for the Rest of Europe*, EUI Working Paper RSC 95/2, Florence: European University Institute.

Eberl, P.C. (1996) *Die Idee des organisationalen Lernens. Konzeptionelle Grundlagen und Gestaltungsmöglichkeiten*, Haupt: Bern.

Grindley, P. (1995) 'Regulation and standards policy: setting standards by committees and markets', in Bishop, M., Kay, J. and Mayer, C. (eds) *The Regulatory Challenge*, Oxford: Oxford University Press, pp. 210–226.

Hakansson, H. (1986) (ed.) *Industrial Technological Development: A Network Approach*, London: Croom Helm.

Heaton, G.R. Jr (1983) 'Regulatory and technological innovation in the chemical industry', *Law and Contemporary Problems* 46 (1), pp. 109ff.

Henning, F. and Klodt, H. (1995) *Wettbewerb und Regulierung in der Telecommunikation*, Tübingen: Mohr.

Irwin, A. and Vergragt, P. (1989) 'Rethinking the relationship between environmental regulation and industrial innovation: the social negotiation of technical change', *Technology Analysis and Strategic Management* 1 (1), pp. 57ff.

Jans, J. (1995) 'The development of EC environmental law', in Winter, G. (ed.), op. cit., pp. 27ff.

Joerges, C. (1996) *Integrating Scientific Expertise into Regulatory Decision-making*, EUI Working Papers RSC 96/10, Florence: European University Institute.

Koch, H.J. (1996) 'Vereinfachung des materiellen Umweltrechts', *Neue Zeitschrift für Verwaltungsrecht* 15 (3), pp. 215ff.

Ladeur, K.H. (1993) 'Risiko und Risikomanagement im Anlagesicherheitsrecht', *Umwelt-und Planungsrecht* 13 (4), pp. 121ff.

Ladeur, K.H. (1995) 'Coping with uncertainty: ecological risks and the proceduralisation of environmental law', in Teubner, G., Farmer, L. and Murphy, D. (eds) *Environmental Law and Ecological Responsibility*, Chichester: Wiley, pp. 299–336.

Ladeur, K.H. (1997) 'European standards and national law', in Joerges, C. *et al.* (eds) *Integrating Scientific Expertise into Standard-Setting*, Baden-Baden: Nomos Verlag.

Luhmann, N. (1988) *Erkenntnis als Construktion*, Berne: Benteli.

Majone, G. (1990) *Deregulation or Re-regulation? Regulatory Reform in Europe and the United States*, London: Pinter Publishers.

Majone, G. (1994) 'The rise of the regulatory state in Europe', *West European Politics* 17, pp. 77ff.

Mitts, A.S. and Holzman, L.R. (1994) 'Products liability and associated perceptions of risk', *Annual Review of Energy and the Environment* 19, pp. 347ff.

Munier, B. (1995) 'Entre rationalité instrumentale et cognitive', *Revue d'Economie Politique* 109 (1), pp. 5ff.

Nelson, R. (1991) 'Why do firms differ, and how does it matter?', *Strategic Management Journal* 12 (1), pp. 61ff.

Noll, R.G. (1982) 'The feasibility of marketable emission permits in the United States', in Stewart, R.B. and Krier, J. (eds) *Environmental Law and Policy*, Charlottesville: Michie, pp. 116ff.

Nonaka, I. (1991) 'The dynamic theory of knowledge generation', *Organization Science* 2 (1), pp. 14ff.

Otway, H. and von Winterfeldt, D. (1992) 'Expert judgement in risk analysis and management', *Risk Analysis* 13 (1), pp. 83ff.

Rajah, N. and Smith, S. (1994) 'Using taxes to price externalities: experiences in Western Europe', *Annual Review of Energy and the Environment* 19, p. 475.

Rasmussen, J. (1995) 'Risk management, adaptation and design for safety', in Sahlin, N.E. and Brehmer, B. (eds), *Future Risks and Risk Management*, Dordrecht: Kluwer, pp. 1ff.

Rehbinder, E. (1995) 'Self-regulation by industry', in Winter, G. (ed.), op. cit. pp. 239ff.

Schlichter, O. (1995) 'Investitionsförderung durch flexible Genehmigungsverfahren', *Deutsches Verwaltungsblatt* 110 (4), pp. 173ff.

Selznick, P. (1985) 'Focusing organisational research on regulation', in Noll, R.G. (ed.) *Regulatory Policy and the Social Sciences*, Berkeley: University of California, pp. 363ff.

Teubner, G. (1995) 'Global Bukowina: the emergence of a transnational legal pluralism', unpublished manuscript.

Wagner, W. (1995) 'The science charade', *Columbia Law Review* 95 (6), pp. 1613ff.

Winter, G. (ed.) (1995) *European Environmental Law*, Aldershot: Dartmouth.

4

ACCESS TO INFORMATION IN A DEREGULATED ENVIRONMENT

Veerle Heyvaert

INTRODUCTION

As the century draws to a close, more and more elements emerge which indicate that the heyday of public administration and bureaucracy, which has so significantly marked governance and public decision-making over previous generations, may belong to the past. Voices are raised to deregulate the state, to trim down administration and to replace the often overbearing and inefficient command-and-control regulations with newer, more flexible and market-oriented methods to implement public policies. The trend towards deregulation has already affected the outlook, procedures and underlying philosophy of many policy areas, including the field of environmental policy.

This chapter aims to link the trend towards deregulation of environmental policies to the movement to greater transparency and availability of environmental information. It suggests that a developed system of access to environmental information could be a useful, even an indispensable mechanism to support deregulatory policies, and to correct some of the shortcomings of deregulation in the area of the environment. To this effect, the chapter provides a brief overview of the roots and development of both the move towards access to information and deregulation, and continues with a discussion of the complementarity and synergism of the two mechanisms. Finally, it asks the question how access to information regimes could be modified or further developed in order better to meet the demands for environmental information in a deregulated society.

OPENING ACCESS TO INFORMATION ON THE ENVIRONMENT

Until the mid-1970s, public administrations in Europe mostly maintained a closed-door policy: the public was not to be informed of internal

55

administrative proceedings, and official documents were not disclosed.[1] However, the last quarter of this century has witnessed a growing trend away from secrecy and towards transparency of administrative proceedings. Scholars have linked this move to the growing complexity of administrative decision-making: the growing volume of the public agenda, the myriad of departments, committees, agencies and administrative working groups that have sprung to life to assist public authorities in the fulfilment of their tasks, and the increasing degree of technicality of the issues administrations deal with, have reduced parliamentary control to a symbolic function rather than a genuine check on executive activities (Erichsen, 1992).

The climate of greater transparency and the development of a right-to-know doctrine also affected environmental policy. In addition to the general concerns regarding the control of administrative decision-making raised above, the demand for access to information on the environment was amplified by a growing awareness of environmental risks. The potential harm to health and the environment that could result from administrative decisions, such as the authorisation of a waste disposal facility, gave the public a stake in these decisions. Clearly, the growing volume of environmental (and consumer) protection legislation and the development of the concept of environmental rights, which in certain countries attained the status of constitutional rights, equally stimulated the public demand for a generalised access to information on the environment.[2]

In Europe, developments of a right of access to information on the environment predominantly took place at the EU level. One of the first documents responding to the demand for information was the 1985 European Parliament Decision on information relating to environmental pollution. It confirmed that information concerning pollution and health risks was often qualified as 'confidential' without reason, and that it remained difficult for citizens to receive adequate information (von Schwanenflügel, 1991). However, the first real landmark was the Fourth Environmental Action Programme drawn up by the European Commission, which, in Section 2.6 on access to information, stresses the importance for individuals to defend their rights and interests, and states that the protection of health and the environment is furthered by access to information (Erichsen, 1992).[3]

In accordance with the principles outlined in the Fourth Action Programme, the Commission started with the preparation of a draft Directive, which was adopted by the Council on 7 June 1990 as Directive 90/313/EEC on the freedom of access to information on the environment (hereafter referred to as the 'Information Directive').[4] The main provisions of this Directive will be studied in detail in the following sections.

However, it is important to remind ourselves that, even if the European level is the most relevant for our further study, there are of course other (international) fora where rights of access to environmental information are being debated and developed. The United Nations are principal actors in this development, as evidenced by, *inter alia*, UNCED's 1992 endorsement of the development of right-to-know legislation in Agenda 21, and the UN/ECE draft Convention on Access to Environmental Information. Also, in the Lugano Convention the Council of Europe included provisions relating to access to environmental information, even expanding the right of access to information held by (private) entities in control of environmentally hazardous activities (Article 16).[5] Finally, efforts are being made to include the right of access to environmental information in the rights protected in the European Convention on Human Rights (ECHR), either through amendment or, more likely, through a new interpretation of existing provisions, in particular Articles 8 (the right to privacy) and 10 (the right to freedom of information) of the ECHR (Weber, 1991).

THE GOALS OF ACCESS TO INFORMATION

In its preamble, the Information Directive points at the problem of the implementation deficit in the area of EU environmental policy. It is hardly a secret that the growing amount of EU legislation aimed at the protection of the environment is not always matched by a similar fervour at the national level, and national governments and administrations are often slow to implement. The preamble continues stating that access to information will result in a better protection of health and the environment. This gives a first indication that the EU legislator shares the conviction that, in the field of environmental protection, more is at stake than the control of the proceedings of public administration. Finally, the preamble points out that harmonisation of access to information requirements is necessary to maintain a competitive level playing field within the internal market. Again, this statement appears implicitly to acknowledge that access to environmental information may have ramifications beyond the relation between public authorities and the public.

As is typical for preambles, the objectives listed remain of a very broad and vague nature (Eifert, 1994). However, several legal scholars have tried their hand at filling in the goals of access to information in a more concrete and detailed manner. In general, they focus on the following two issues: (1) access to information as a means of control of public authorities, and (2) as an alternative regulatory mechanism (Winter, 1990).

Control of public authorities

Daniel Alexander (1990) called the Information Directive 'an effort to devolve power away from public authorities to interested citizens and interest groups'. In the same vein, von Schwanenflügel (1993) confirmed that the right to exclusive decisional discretion over information refers to power, and thus a public right of access to a diminishment thereof, and Schertzberg (1994) labelled the increased involvement of private citizens and interest groups in the goings-on of the executive a form of decentralisation and privatisation of executive controls. These comments all echo a shift from notions of popular representation to popular sovereignty, and the case for a more direct review of administrative activities than through parliamentary control.

Yet, the discussions on the control of public authorities already suggest that, in the area of environmental policies, more is at stake than a straightforward check on the executive powers. Thus, after making reference to the shift of power, von Schwanenflügel continues that the Information Directive offers citizens an opportunity to control polluters, thereby assisting the public authorities in the exercise of their duties. Similarly, Morand-Deviller (1992) links the interest that the public has in administrative decisions concerning environmental protection and urban planning, to a public task as a controller and helper of administration.[6] In these descriptions, access to information appears not so much to target the activities of administrations, but rather those of polluters.

Access to information as an alternative regulatory mechanism

Whereas the first objective of access to environmental information focuses rather on the control of the procedures for implementing environmental policy, the second objective is related more closely to the content of the information itself. In this respect, public access to environmental information is said to correct the failure of the market to supply such information at a socially optimal level. Environmental quality being a collective good, the demand for pollution control will usually fall short of the desirable level, and access to information aims to strengthen this demand by sensibilising and mobilising the public (Huppes and Kagan, 1989).

The informed citizen may contribute to the improvement of environmental quality either through the political process, demanding higher standards of protection, more stringent regulation and swifter implementation (a voice strategy), or through the market process, giving preference to products produced by 'green' industries and shunning those produced by polluting industries (an exit strategy) (von Schwanenflügel, 1993; Pease, 1991; and Eifert, 1994). As the demand for environmental quality rises, and more information on company behaviour becomes available through public access, industry will be inclined to improve its environmental track-record (van Gestel, 1994).

Finally, Eifert (1994) points at the benefits which greater transparency and increased communication may bring for the assessment and evaluation of environmental risks. It is widely acknowledged that public perception of risks frequently varies to a great extent from industry's perception, and both views may differ considerably from those upheld by administration (Schrader-Frechette, 1991; Johnson, Sandman and Miller, 1992; Wagner, 1995). Exchange of information between the different players might prove a useful tool to approximate their views, and ultimately result in a reformulation of the public agenda, more responsive to the concerns of both industry and the public.

In conclusion, access to environmental information has been hailed as a strategy to complement, or even to supplant traditional command-and-control regulation (Pease, 1991). Whereas the latter directly imposes binding environmental standards and requirements on its addressees (usually industry), and leaves little room for dialogue, access to information operates in a more indirect, cooperative and – so certain authors claim – flexible manner, relying on a combination of regulatory and market impulses to attain a higher degree of environmental protection (von Schwanenflügel, 1993). Thus, the call for greater transparency shares many of the underlying considerations and characteristics of the deregulation movement.

DEREGULATION

The lack of transparency is one, but by no means the only aspect of administration which has come under attack. The last decades are characterised by a growing discontent with the heavy machinery of bureaucracy, and a corresponding belief that certain public functions could be performed less costly and more effectively by non-governmental entities (Lüder, 1996; Grabovsky, 1994). Amato (1985) speaks of a 'rediscovery of the market': in order to obtain a higher level of efficiency and to develop a more flexible economy, it is preferable to let the market run its course, rather than burden the economy with a battery of detailed rules and prescriptions. Deregulation aims to avoid ineffective regulation and narrow down the regulatory scope, to limit interventions which adversely affect market competition, and to make laws more simple, manageable, comprehensive and flexible (Stober, 1995).

In practice, deregulation usually refers to one of the two following activities: the privatisation and opening up to competition of state monopolies, or the reformulation and narrowing down of the regulatory scope in policy areas, such as industrial, environmental and consumer policy. It is deregulation in the latter sense which forms the focus of our study.

The latter type of deregulation efforts frequently entail a redefinition of administrative functions rather than their abolishment. Instead of drawing up hundreds of detailed rules and requirements, the administration in a deregulated environment would impose and/or watch over the compliance with broad framework conditions and essential requirements, leaving their implementation up to the regulated parties (van Gestel, 1994). Because of its emphasis on redefinition of functions and reallocation of responsibilities, some prefer to label this trend as 're-regulation' or 'responsive regulation' (Ayres and Braithwaite, 1992). The 1985 EC new approach to harmonisation of product regulations is an example of this broad view of deregulation: rather than detailing the various requirements which products have to comply with in order freely to circulate on the common market, EC harmonisation Directives limit themselves to enumerating essential safety requirements, and rely on non-governmental standardisation bodies (CEN, CENELEC and ETSI) to elaborate specific product-, design- and process-standards which are presumed to satisfy the essential requirements.[7] This form of deregulation is also referred to as self-regulation, because it is up to the regulated party (*in casu*, industry) to detail its own constraints (Baram and McAllister, 1982).

In environmental policy, exercises in deregulation are often based on ideas of cooperation. For example, the Dutch packaging covenants and standards on environmental management, such as the UK standard BS 7750, are created and maintained through cooperation between different industrial enterprises (Christie, Rolfe and Legard, 1995; Salter, 1994). Eco-labelling, on the other hand, relies on mechanisms of communication concerning product characteristics and consumer preferences between industry and consumers (Grabovsky, 1994). Finally, certain forms of internalisation of environmental management, such as the EU eco-management and audit scheme (the 'EMAS Regulation'), involve intricate distributions of tasks and exchange of information between public authorities and private industry (Sellner and Schnutenhaus, 1993; Winter, 1994).[8]

So far, attempts at deregulation of environmental policy in Europe have been tried out on a fairly limited scale and, particularly in heavily regulated countries such as Germany, still rest on a broad regulatory framework. However, if these efforts are successful, the scope of deregulation in environmental policy may well expand in the future. Consequently, it becomes indispensable to go beyond discussions on the real benefits of deregulation schemes which have already been carried through on a limited scale, and the potential benefits of deregulation implemented on a large scale, and turn our attention to the other side of the coin: if we continue to deregulate environmental policies, what are the potential weaknesses of and limitations to the deregulatory movement, and how can they be resolved and amended? The following sections offer a first attempt to address some of the problems of deregulating the environment.

Challenges to deregulation

Deregulation of environmental protection policies is, to some extent, a paradoxical concept. On the one hand, proponents of deregulation and a stronger reliance on market forces contend that such mechanisms allow a more efficient management of the environment, and can result in a higher degree of protection than guaranteed under traditional regulatory schemes (Ackermann and Stewart, 1985). Furthermore, in light of the 'polluter pays principle', it seems logical that the care as well as the costs of environmental protection should be allocated to the polluters (Xiberta, 1994). On the other hand, granting more decisional power concerning environmental protection and management to those parties which are also its main assailants, appears to require a considerable leap of faith. Also, the growing public expectation to be protected against environmental harm, and the increasing importance political parties have attributed to environmental issues during the last decades, seem at odds with a decreasing role of government and administration in the pursuit of higher environmental standards. If deregulation continues to develop as a major trend in environmental policy, it will have to confront the obstacles of public duty, market failure and trust.

Constitutional or legal public duties conferring the task of the care for the environment to public authorities set a minimum requirement for public activity on behalf of the environment and, correspondingly, a ceiling to the maximum allowable degree of deregulation (van Gestel, 1994). Examples of such legally enshrined public duties can be found in the Dutch Constitution. At the EU level, Article 130 of the Treaty on the European Union elevates environmental protection as one of the goals and tasks of the EU. At the international level, a doctrine of the existence of a positive duty of the state to ensure environmental protection is gradually being developed under the auspices of the European Court of Human Rights. Obviously, a duty for public authorities to act does not necessarily imply that this task has to be filled in by means of command-and-control regulation. Nevertheless, it does mean that public authorities have to be enabled to perform functions which contribute to environmental protection in a meaningful way.

The obstacle of market failure in a deregulated environment is linked to a lack of internalisation of economic costs: industrial production decisions are not capable of reflecting the aggregate costs and benefits of environmental damage. It is often extremely difficult to determine causal links between industrial pollution and environmental effects, and to measure the amount of harm created (Applegate, 1991). Furthermore, even where accurate data on causation and on the seriousness of expected harm are available, the conversion of these data into measurable economic units of cost remains an extremely complex and controversial undertaking (Pearce, 1976). Finally, when regulatory constraints are lifted, we may also question the willingness of industry to generate data which in all likelihood will cast its activities in a negative light (Lyndon, 1987). Again, the obstacle of market failure does not necessarily mandate a complete change of course and a return to command-and-control regulation. However, it is essential that any environment, including a deregulated one, provides the necessary incentives or constraints to 'fill in the environmental data gap'.

The last challenge to deregulation to be addressed in this context goes back to the paradox of handing over the care of the environment to its main assailants. Although many industrial sectors have made a concerted effort to improve their environmental reputation, the images of 'green crusader' and 'corporate raider' still do not blend easily. Moreover, environmental interest groups and, more generally, the public very probably have sound reasons to be distrustful or at least sceptical of industry as the self-proclaimed guardian of environmental quality. For example, while offering some improvements to environmental protection, voluntary agreements may equally be interpreted as pre-emptive strikes to avoid imminent and more stringent regulation, enabling industry to set the terms and simultaneously to demarcate the limits of environmental protection. Also, whereas industry purports to be a strong supporter of internalised eco-management and audit schemes, it is remarkable that during the drafting stages of the EMAS Regulation the industrial lobby vehemently (and successfully) opposed any suggestions to make EMAS mandatory (Xiberta, 1994). Finally, although certain public institutions – such as the European Commission – are quick to claim that cleaner technologies will ultimately be less costly and/or more profitable for industry, there are economic studies which indicate that such statements, pronounced in a generalised manner, may be far too optimistic.

It is not our intention to assess to what extent industry can be trusted with the care of the environment. However, it should be noted that many factors, including past experiences, the reluctance of industry to assume any binding obligations, and the potential conflicts between environmental quality and economic profitability, complicate this trust.

Access to information in a deregulated environment

The challenges and obstacles to deregulation of environmental policies described above could be confronted and – at least partially – overcome through the establishment of a general and sophisticated access to an environmental information system. The conceptual similarities between access to information and deregulation are striking: both ideas were developed roughly during the same time period, and both arise to a considerable extent from a growing dissatisfaction with the traditional and non-communicative forms of governance and administration. In its stead, access to information as well as deregulation propose systems of organisation which are more cooperative, flexible, and replace the top-down with a bottom-up approach.

What sets the two trends apart are the different interest groups supporting access to information on the one hand, and deregulation on the other. Environmental interest groups and the public in general have reasons to distrust industry as the protector of the environment, and consequently to distrust deregulation. Industrial enterprises, in turn, have cause to be wary of free access to information: it may cast a negative image of industry, and, because of the sensitivity of certain industry-related data, may put the disclosing company at a competitive disadvantage (Turieux, 1994).

Because of these conceptual similarities, in combination with the different agendas of the interest groups advocating the two trends, access to information is a particularly apt system to overcome the weaknesses and shortcomings of deregulation, without having to relinquish deregulation's underlying ideology of market-orientation, flexibility and decentralised decision-making. First, it provides a basis for trust. It needs little explanation that the fear of being cheated reduces considerably when all players have their cards on the table. Second, because of its awareness-heightening and mobilising effects on consumers, access to information contributes to correcting the market failure caused by a lack of internalisation of costs (Turieux, 1994). Admittedly, access alone will not solve the entire information deficit, since it predominantly targets disclosure of existing information, but does not directly compel companies to create more information. In order to remedy the information gap in a more effective manner, it would therefore be advisable to expand industry's supply of environmental information duties to include research, monitoring and reporting duties. Third, the access to information system also supplies public authorities with a new, meaningful task within a deregulated framework. Facilitating, supervising and controlling the flow of environmental information, deciding on the confidentiality of industrial data and settling disputes concerning supply as well as access to information, could be interpreted as public duties which meet the legal minimum requirements for public involvement on behalf of the environment.

The conclusion that deregulation and access to environmental information should go hand-in-hand is by no means a shocking one. Already, EC instruments which foresee deregulatory and/or internalised forms of environmental management, such as the EMAS Regulation and the impending revision of the Seveso Directive, contain requirements to make information on the subjects at issue available to the public. Xiberta (1994) furthermore pointed out that: '[T]he contents of the (Information) Directive . . . will complement the eco-management and audit scheme, since the easiest way in which truthful information on environmental performance can be held by public authorities is by means of submission of the environmental reports from eco-audits.' However, we reiterate that, if deregulatory policies prove successful and their application on a large scale is contemplated, it becomes necessary to question whether the existing access to environmental information schemes are sufficient and adequate to fulfil their complementary functions in a deregulated environment, or whether changes will need to be made.

THE ORGANISATION OF ACCESS TO INFORMATION IN A DEREGULATED ENVIRONMENT

The organisation of access to environmental information in a deregulated framework entails a vast amount of issues, problems and structural options. They range from specific organisational issues, such as whether a price should be charged for the supply of information and how to determine a reasonable price, to very broad questions, including legislative and judicial protection of the right of access, and the relation between access to information and access to justice. An exhaustive discussion of all issues involved would exceed the scope of this paper. Nevertheless, in a preliminary attempt to provide an overview of some of the basic questions concerned, the following sections address three of the key issues relating to the organisation of access to information: the institutional organisation, the scope of information duties, and the treatment of trade secrets.

The institutional organisation of access to information

Who should be responsible for the supply of environmental information to the public? The Information Directive points in the direction of public authorities: according to Articles 3(1) in combination with 2(b), any public administration at national, regional or local level with responsibilities relating to the environment, and possessing information, is required to make this information available.[9] However, in a context of deregulation, with responsibilities for the environment shifted towards industry, the question arises whether private organisations, such as companies with internalised environmental management systems, should not be included within the scope of the Directive. The Information Directive as it stands gives an indication that there is room for enlargement: Article 6 provides that information relating to the environment should equally be supplied by 'bodies with public responsibilities for the environment and under the control of public authorities'. The meaning of the terms 'public responsibility' and 'control of public authorities' have been the subject of much scholarly debate; however, there appears to be a consensus that the Directive only targets bodies which have received a specific public mandate to deal with environmental issues (Turieux, 1994; Schertzberg, 1994; Erichsen, 1992). Also, the Department of Environment explanatory Annexe to the UK Environmental Information Regulation 1992 (which implements the Information Directive) supplies a long list of relevant bodies with responsibilities for the environment, which does not include one private organisation (Birtles, 1993). In conclusion, there are no interpretations thus far which include private bodies within the scope of the Information Directive.

The EMAS Regulation, on the other hand, does impose a direct duty on its participating companies to disseminate the validated environmental statement to the public. However, this duty to communicate information to the public is restricted to the environmental statement, and thus falls short of conveying a general right of public access to information *vis-à-vis* private parties with responsibilities for the environment (Sellner and Schnutenhaus, 1993). The proposal to amend the Seveso Directive – which is not a deregulatory measure in a strict sense but does impose a far-reaching system of internalised environmental management – establishes an active duty to inform persons liable to be affected by a major accident on safety measures and on the requisite behaviour to adopt in the event of an accident. Moreover, this information is to be made permanently available to the public (Barrat and Enmarch-Williams, 1994). We can infer from Article 13(4) that in some instances, the companies are personally responsible for making the information, contained in a 'safety report', available to the public. However, this situation has only been specified for exceptional cases, and it remains unclear whether the provision should be interpreted to cover all safety reports.

The sparse indications provided by the existing EU legal framework thus far do not allow us to draw general conclusions with regard to the institutional organisation of access to information in a deregulated environment. Therefore, we will attempt to answer the question 'how should access to information be organised' by reviewing the pros and cons of two alternative arrangements: access to information provided by public authorities; and access to information guaranteed by private parties.

The benefits of making private parties responsible for the availability of environmental information, are in the first place that this arrangement would be in line with the 'polluter pays' principle, since the burden of gathering, organising and disseminating the information would fall on industry's shoulders, and that it would place information duties in the hands of those parties which are most knowledgeable about the data concerned. Making private companies directly responsible for public access to information would additionally unburden public administration, which could be a further step towards an encompassing deregulated, market-oriented framework. Yet, there are also serious disadvantages linked to a private organisation of public access to information.

First, it would lead to a fragmentation of environmental information being held by a range of different companies and institutions. According to Eifert (1994), the current system, where information is primarily supplied by public authorities, already faces criticism because the information is scattered over different administrative departments. This complicates efforts to acquire a general overview of the state of the environment and threats emanating from different sources. It is precisely this type of 'general overview information' that is the most useful for public opinion formation (Eifert, 1994). Moving towards a system of access to information held by private companies would exacerbate this situation. Second, the privatisation of access to information could result in the implementation of many different forms of access and procedures, which might result in serious discrepancies in the accessibility of information.[10] Third, direct information duties between industry and the public will necessitate a considerable conversion and modification of company practices. It remains to be seen whether such an investment into information services is feasible for all parties involved. Finally, the problem of trust re-emerges: when the supply of information becomes a private responsibility, so does the determination whether data are confidential, commercially sensitive and thus exempted from public access. Considering industry's reluctance to disseminate potentially damaging information relating to its own activities, it may be desirable to leave such determinations up to a third, less partial, party.

It is not unlikely that, if given the responsibility for access to information, industry would set up organisations specifically dedicated to the supply of information to the public. The establishment of such organisations, which we might imagine as some type of private counterparts to the European Environmental Agency, could resolve many of the problems listed above, since these bodies would centralise data, would in all probability establish uniform or at least fixed sets of procedures for access to information, and would limit the adaptations required within individual companies (Alexander, 1990). The trust problem, however, remains.

This last problem could be resolved (or at least mitigated) by returning responsibilities for access to information to public authorities. Public authorities provide an external check on industry's claims relating to the sensitivity of the data submitted, which increases the reliability and trustworthiness of industry's assessments. A further reason to leave the responsibility for access to information with public authorities, is that, to a significant extent, the 'infrastructure' is already present. During these last years, public authorities have developed procedures to deal with information requests and exchange of information in general, established networks or institutions whose main tasks concern information and communication (the European Environmental Agency; the CORINE and RAPEX information exchange programmes), made arrangements for supervision and review of administrative decisions relating to access, and accumulated a considerable amount of experience. For example, the French administration is assisted by an advisory body, the Commission d'accès aux documents administratifs, which was established in 1978 to give opinions on administrative decisions relating to the public release of information (Letterton, 1995). In other instances, pre-existing bodies have assumed new responsibilities covering data-gathering as well as disclosure (Huppes and Kagan, 1989).

Because of the need for control of industry's assessments and the advantages of existing arrangements and experience, it would be advisable to maintain public authorities' involvement in access to environmental information. However, there is certainly room for improvement. We have already indicated the problem of information being scattered over different administrative bodies. The designation of a specific body which deals with information requests, re-roots these requests to the departments concerned, and then gathers and (possibly) processes the replies to forward them to the requesting party, could be one way of organising a more centralised system of access to information. Additionally, there is the problem that public authorities usually have less expertise concerning the information and its significance than the industrial experts which submit the information. Here, some form of public–private participation might bring a solution, either through the creation of 'mixed bodies', or through the development of requirements on industry to make information user-friendly.[11] In light of the risk of capture that the first alternative entails, the second seems preferable, and will be developed in more detail in the following section.

The scope of information duties

The second basic question affects the scope of information duties: which efforts do public authorities and industry have to make in order to ensure the availability of environmental information? Going back to the Information Directive, the information duties imposed on public authorities are rather 'passive' in nature: public authorities are only to grant access to information which is readily at their disposal; they do not have to gather, collect or process additional information at the information-seeking party's request (Schertzberg, 1994). Environmental information at the disposal of public authorities has usually been obtained either in compliance with legal and regulatory requirements (e.g. notification of new dangerous substances, application for authorisation or licensing) or as a result of monitoring and sampling activities undertaken by the authorities or on their behalf (Turieux, 1994). Furthermore, it should be noted that the Information Directive does not specify how environmental information should be made available; access can be granted, *inter alia*, through disclosure of the original documents, through the dissemination of copies, or even by means of information given orally. Thus, public authorities' information duties pursuant to the Directive are relatively narrow, and not very strictly defined.

In an attempt to give a broader interpretation to public duties to ensure access to information, certain scholars have stressed the changing circumstances due to new and developing means of communication: the existence of central databases and networks reduces the act of gathering information – irrespective of its source – to an extremely simple operation; an effort not bigger than retrieving in-house documents. According to this line of reasoning, public information duties should include gathering activities, if it involves no extra effort (Turieux, 1994). Others refer to the *effet utile* doctrine. In the context of access to information, *effet utile* could be interpreted to mean that access to information should equal access to the original documents, or at least a certified copy (Schertzberg, 1994). Access to information furthermore loses its utility if the documents are illegible or incomprehensible to the requesting parties. Here, *effet utile* may even be used to oblige public authorities to make sure that the information is really accessible (Erichsen, 1992).

In light of the goals of access to information in a deregulated environment, particularly those of bridging the information gap and establishing trust between the public and polluters, the concern for 'real' accessibility and comprehension of environmental information would grow even stronger in the case of deregulation. This pleads for a broader and more precise definition of public authorities' information duties than presently provided in the Information Directive. However, two considerations should be taken into account. First, we recall that the deregulatory movement was to a considerable extent a reaction to administrative inefficiency. Reforms of administration should therefore be considered carefully, so that old burdens are not merely replaced by newer and heavier ones. Second, we repeat that the companies producing and submitting environmental information are generally more knowledgeable about the data concerned than public authorities. Consequently, we should equally consider industry's role in the production of user-friendly environmental data to facilitate 'real' access to information.

As discussed above, companies submit information to public authorities pursuant to legal and regulatory requirements. However, whereas the information relating to the environment used to be primarily a means to an end, the objective being, e.g., the approval of a licence, authorisation or permit, the submission of information on the environment more and more becomes an end in itself, only very indirectly related to ensuing administrative decisions. Here, we speak of 'environmental reporting duties' (van Gestel, 1994). The establishment and/or further development of environmental reporting duties would form a considerable contribution to the correction of the environmental information deficit, since it does not only oblige companies to make existing information public, but can equally ensure that environmental information is created in the first place (Padgett, 1992). Turieux (1994) has suggested that all large companies should consider the introduction of an environmental information management system. With respect to the creation of new data, reporting duties could be supplemented by monitoring duties, which would enable regular updates or amendments to the information originally submitted. Additionally, reporting duties may include requirements to make the information understandable to the public. Such a requirement is already included with respect to the environmental statement submitted by companies which adopt the EMAS scheme.[12] The reinforcement of environmental reporting (and monitoring) duties thus appears a particularly appropriate manner to complement deregulation schemes, maintaining a reasonable degree of control, however, without unduly burdening public administration (Turieux, 1994).

The treatment of trade secrets

The last and possibly most difficult issue to be dealt with within the confines of this chapter, is the treatment of commercial and professional secrets (hereinafter generically referred to as 'trade secrets') in access to information systems in a deregulated environment. The issue of trade secrets and their protection presents the conflict between companies' interests to protect their competitive position and the public's interest in obtaining information in its starkest form. It stands beyond dispute that information represents a high commercial value, and that unwarranted disclosure may signify a severe commercial setback and disruption. On the other hand, censored versions of industrial information relating to the environment may be so diluted that they lose any practical value for the public. Information relating to pesticides is a case in point: research and risk assessment of new pesticides is so enormously costly, that full publication of the research data would probably lead to a cessation of the production of new pesticides. Yet, the sanitised versions of research reports are probably insufficient to allow the public to make an informed judgement concerning the potential health impacts of the product (Jans, 1990).

In all probability, the question of the degree of protection which should be granted to trade secrets will pose itself all the more urgently in a deregulated environment. Where the implementation of environmental programmes is left more and more to the discretion of industrial companies, the scope of trade secrets potentially will multiply, since the development and in-company application of environmental management programmes represent an economic value to the implementing company. Therefore, when considering the large-scale application of deregulatory policies, it becomes imperative to study possible, equitable arrangements for the protection of trade secrets.

At present, all EU legal and regulatory instruments containing provisions on the submission or disclosure of information by commercial enterprises include a trade secrets exemption. However, it is remarkable that 'trade secrets' are nowhere positively defined. For example, the Information Directive provides: '[M]ember States *may provide* for a request for such information to be refused where it affects: . . . commercial and industrial confidentiality, including intellectual property' (emphasis added).[13] It supplies however no explanation or examples of what constitutes this confidentiality. An overview of national interpretations of this clause quickly leads to the conclusion that the views held by Member States on the protection of trade secrets, and on the relevant clause in the Information Directive, differ considerably.

The French author Roseline Letteron (1995) reads the Directive's trade secrecy clause in combination with Article 214 of the EC Treaty, which covers the obligation of professional secrecy of Community officials and includes 'information relating to undertakings and their commercial relations' in the realm of confidentiality. She arrives at the conclusion that the stipulations of the Information Directive are therefore largely superfluous, and that all information transmitted by private enterprises is covered by rules relating to professional secrets. We may question the validity of this assessment. First, Article 214 of the EC Treaty targets Community officials and, although it may serve as a guideline, does not bind national administrations. Second, reading Article 214 in such a literal and rigid manner points towards a blindness to recent developments at the EU level, which have mitigated the severity of professional secrecy through the adoption of rules and decisions introducing a greater degree of transparency and openness of EU institutions.

Turning towards the Netherlands, we perceive a slightly more nuanced interpretation of trade secrets. For example, the Dutch Parliament's reading of trade secrets in the Dangerous Substances Law restricts trade secrets to '[t]hose data publication of which could harm the competitive position of the company' (translation from Dutch) (Jans, 1990). This definition introduces a criterion of harm into the concept of trade secrets, which restricts its scope. However, the narrow reading by the Dutch Parliament has been largely undone by Dutch case law, which uses a broader interpretation: if it is possible to infer from data information 'worth knowing' relating to the technical exploitation, production processes, the sale of products, the clientele or the suppliers of commercial enterprises, these data are covered by commercial secrecy.[14]

A particularly narrow reading of trade secrets can be found in the UK Governmental Guidance note concerning the Environmental Information Regulations of 1992. It provides that: '[W]hen appropriate, the relevant person can decide on the merits of the evidence whether the release of identified information would prejudice the supplier's commercial interests . . . it will *not normally* be appropriate to withhold information in response to a general claim that disclosure might damage the reputation of the supplier and hence his commercial competition. Where it is agreed that information should be withheld, this should be limited to the minimum time necessary to safeguard the commercial interest' (emphasis added) (Salter, 1994). Here, we perceive the first indications that a balancing exercise should be performed: in order to determine whether information should be withheld from the public, the evidence must be weighted. The need for a balancing test is further developed in German jurisprudence. According to Knemeyer (1993), not all commercial secrets have an absolute character; there is room for evaluation. This explains why the Information Directive explicitly stipulates that access *may* be refused for reasons of commercial secrecy, thus implying that, in certain cases, Member States may (or even should) decide not to deny access for these reasons. In this regard, it is interesting to mention a recent decision by the European Court of First

Instance relating to public access to Council documents.[15] Here, the Court gave its interpretation of Article 4(2) of Decision 93/731/EC on public access to Council documents, which provides: '[A]ccess to a Council document may be refused in order to protect the confidentiality of the Council's proceedings' (p. 3). The Court took this 'may be refused' to mean that a balancing test *must be* performed to weigh the individual's interests in obtaining access *versus* the interest of the Council in retaining confidentiality (Sprokkereef, 1996). Admittedly, this case deals with Council documents and not trade secrets. However, the similarities are striking.

A decision to withhold information must furthermore pass the proportionality test, being the least restrictive measure possible to achieve the desired goal. Thus the Information Directive states that: '[I]nformation . . . shall be supplied in part where it is possible to separate out information on the items concerning the interests referred to above (including commercial and industrial confidentiality).'[16] The proportionality principle also resonates in the UK Government's requirement that 'this (non-disclosure of information) should be limited to the minimum time necessary'. In Germany, Erichsen (1992) equally mentions a condition that the evaluation of trade secrets should take into account the principle of '*Verhaltnismäßigkeit*' (proportionality) (Erichsen, 1992). However, national differences may still persist. For example, in Germany the interests of industry are expressed in terms of their right to self-determination, whereas in the UK, the emphasis is on non-distortion of the undertaking's competitive position. A different view on the objectives of confidentiality may well lead to a different understanding of which measures are proportionate, and which exaggerated.

In conclusion, on the basis of existing literature we are able to trace the contours of a rudimentary framework and identify some recurring criteria of a regime for the protection of trade secrets. However, the information is sketchy at best, and there appears to be a number of discrepancies between different EU Member States. In a deregulated environment, where a lion's share of the information relating to the environment will be generated by commercial enterprises and problems of commercial confidentiality will become even more prevalent than they are today, the substantial and procedural development and fine-tuning of a flexible and proportionate system to deal with trade secrets should become a priority. A detailed study should not only cover the criteria to determine the confidentiality of different categories of commercial and industrial data, but will ultimately have to address issues of organisation (in other words, who should have the first, and who the final words in such a determination) and oversight. Moreover, if the European Union wishes to adhere to its principle of 'creating a level playing field', and considering the distorting effects different national confidentiality regimes may have on competition, this effort should probably be undertaken at the EU level.

CONCLUSION

By linking deregulation to access to environmental information, this chapter has attempted to propose a system which meets the demands of market-orientation and economic efficiency, while maintaining a reasonable degree of public control. The guarantee of transparency, not only of public authorities' but also of industry's behaviour, may provide a solid basis for the still insufficient and much-needed trust between the public and commercial enterprises in their implementation of deregulated environmental programmes and policies. However, in order to attain the dual goal of efficiency and transparency, many improvements are still to be made. They range from the institutional organisation of access to information to the development and harmonisation of a reliable and reasonable regime for the treatment of trade secrets. Needless to say, the implementation of a sophisticated system of access to information will require a significant investment, and this should be taken into account when deregulatory policies are being considered. The alternative, to push deregulation of the environmental sector forward without corresponding efforts to improve access to environmental information (or to establish a different regime able to cope with the identified shortcomings of deregulation), might result in a sacrifice of legitimacy on the altar of efficiency. This analysis does not aim to deter policy-makers from considering deregulation in the area of the environment, but merely draws attention to the efforts and changes which will be required in order to achieve 'good deregulation'. Or, to paraphrase a Dutch author, deregulation does not implement itself.

NOTES

1 A notable exception to this regime was the Swedish 'Tryfriheitsverordnung', which established a general right of access to information as of the end of the nineteenth century (Gurlit, 1989).
2 See e.g. Article 21 of the Dutch constitution laying down a public duty to protect the environment.
3 Fourth Environmental Action Programme, O.J. C 70/3 (1987).
4 O.J. L 158/56 (1990).
5 Treaty concerning Civil Liability for Damage caused by Activities Dangerous to the Environment, 21 June 1993.
6 See also Eifert (1994) on the participatory nature of informed cooperation by citizens in the public process.
7 Council Resolution of 7 May 1985 on a new approach to technical harmonization and standards, O.J. C 136/1 (1985).
8 Council Regulation (EEC) No. 1836/93 of 29 June 1993 allowing voluntary participation by companies in the industrial sector in a Community eco-management and audit scheme, O.J. L 168/1 (1993).
9 With the exception of public bodies acting in a judicial or legal capacity (Article 2(b)).

10 Depending, e.g., on the industrial sector (or even the individual company) or the type of information involved.
11 The EMAS Regulation already contains such a requirement in Article 5(2): '[T]he environmental statement shall be designed for the public and written in a concise, comprehensible form. Technical material may be appended.'
12 Article 5(2) of the EMAS Regulation.
13 Article 3(2), 5th indent of the Information Directive.
14 Vz. ARRvS, 3 Dec. 1984, AB 1986, 54.
15 Case T-194/94: *John Carvel and Guardian Newspapers Ltd v Council of the European Union*, 19 October 1995 (not yet reported).
16 Article 3(2), last indent.

REFERENCES

Ackerman, B.A. and Stewart, R.B. (1985) 'Reforming environmental law', *Stanford Law Review*, pp. 1333–1365.
Alexander, D. (1990) 'Freedom of access to information on the environment', *New Law Journal*, 140 (6472), pp. 1315–1316.
Amato, G. (1985) 'Problemi di governo e della deregulation', in Cassese, S. and Gerelli, E. (eds) *Deregulation. La Deregolamentazione Amministrativa e Legislativa*, Milano: Franco Angeli.
Applegate, J.S. (1991) 'The perils of unreasonable risk: information, regulatory policy, and toxic substances control', *Columbia Law Review*, 91(2), pp. 261–333.
Ayres, I. and Braithwaite, J. (1992) *Responsive Regulation. Transcending the Deregulation Debate*, Oxford: Oxford University Press.
Baram, M.S. and McAllister, K. (1982) *Alternatives to Regulation*, Massachusetts: Lexington Books.
Barrat, R. and Enmarch-Williams, H. (1994) 'Major industrial accident hazards and the proposed new "Seveso" Directive', *European Environmental Law*, 3 (7), pp. 195–199.
Birtles, W. (1993) 'A right to know: the environmental information regulations 1992', *Journal of Planning and Environmental Law*, pp. 615–625.
Christie, I. and Rolfe, H. with Legard, R. (1995) *Cleaner Production in Industry*, London: PSI Publishing.
Eifert, M. (1994) 'Umweltinformation als Regelungsinstrument', *Die Öffentliche Verwaltung*, 13, pp. 544–552.
Erichsen, H.-U. (1992) 'Das Recht auf freien Zugang zu Informationen über die Umwelt', *Neue Zeitschrift für Verwaltungsrecht*, 5, pp. 409–419.
Grabovsky, P.N. (1994) 'Green markets: environmental regulation by the private sector', *Law & Policy*, 4, pp. 421–422.
Gurlit, E. (1989) 'Europa auf dem Weg zur gläsernen Verwaltung?' *Zeitschrift für Rechtspolitik*, 7, pp. 253–257.
Huppes, G. and Kagan, R.A. (1989) 'Market-oriented regulation of environmental problems in the Netherlands', *Law & Policy*, 2, pp. 215–239.
Jans, J.H. (1990) 'Passieve openbaarheid in het milieurecht', *Tijdschrift voor Milieu Recht*, 4, pp. 146–157.
Johnson, B.B., Sandman, P.M. and Miller, P. (1992) 'Testing the role of technical information in public risk perception', *RISK – Issues in Health & Safety*, 3, Fall, pp. 341–359.
Knemeyer, F.-L. (1993) 'Die Wahrung von Betriebs- und Geschäftsgeheimnissen bei behördlichen Umweltinformationen', *Der Betrieb*, 14, pp. 721–724.
Letterton, R. (1995) 'Le modèle français de transparence administrative à l'épreuve du droit communautaire', *Revue Français de Droit Administratif*, 1, pp. 721–726.

Lüder, K. (1996) 'Triumph des Marktes im öffentlichen Sektor?' *Die Öffentliche Verwaltung*, 3, pp. 93–100.

Lyndon, M.L. (1987) 'Information economics and chemical toxicity: designing laws to produce and use data', *Michigan Law Review*, 87(7), pp. 1794–1861.

Morand-Deviller, J. (1992) 'Les instruments juridiques de la participation et de la contestation des décisions d'aménagement', *Revue Juridique de l'Environnement*, 4, pp. 453–467.

Padgett, M. (1992) 'Environmental health and safety – international standardization of right-to-know legislation in response to refusal of United States multinationals to publish toxic emissions data for their United Kingdom facilities', *Georgia Journal of International and Comparative Law*, 22, pp. 701–717.

Pearce, D. (1976) 'The limits of cost–benefit analysis as a guide to environmental policy', *Kyklos*, 29(1), p. 97–112.

Pease, W.S. (1991) 'Chemical hazards and the public's right to know. How effective is California's Proposition 65?', *Environment*, 10, pp. 11–20.

Sadofsky, D. (1990) *Knowledge as Power. Political and Legal Control of Information*, New York: Praeger Publishers.

Salter, J. (1994) 'Environmental information and confidentiality concerns', *European Environmental Law Review*, 3(10), pp. 289–293.

Schertzberg, A. (1994) 'Freedom of information – deutsch gewended: Das neue Umweltinformationsgesetz', *Deutsches Verwaltungsblatt*, 109, 1 July, pp. 733–745.

Schrader-Frechette, K.S. (1991) *Risk and Rationality. Philosophical Foundations for Populist Reforms*, Berkeley and Los Angeles: University of California Press.

Sellner, D. and Schnutenhaus, J. (1993) 'Umweltmanagement und Umweltbetriebsprüfung ('Umwelt-Audit') – ein wirksames, nicht ordnungsrechtliches System des betrieblichen Umweltschutzes?', *Neue Zeitschrift für Verwaltungsrecht*, 10, pp. 928–934.

Sprokkereef, A. (1996) 'European Court of Justice: case report', *European Environmental Law Review*, 5(1), pp. 23–28.

Stober, R. (1995) 'Deregulierung im Wirtschaftsverwaltungsrecht', *Die Öffentliche Verwaltung*, 4, pp. 125–135.

Turieux, A. (1994) 'Das neue Umweltinformationsgesetz', *Neue Juristische Wochenschrift*, 36, pp. 2319–2324.

Van Gestel, R.A.J. (1994) 'Zelfregulering door bedrijfsinterne milieuzorg gaat niet vanzelf', *Tijdschrift voor Milieu en Recht*, 6, pp. 166–175.

Von Schwanenflügel, M. (1991) 'Das Öffentlichkeitsprinzip des EG-Umweltrechts', *Deutsches Verwaltungsblatt*, 106, 15 January, pp. 93–101.

Von Schwanenflügel, M. (1993) 'Die Richtlinie über den freien Zugang zu Umweltinformationen als Chance für den Umweltschutz', *Die Öffentliche Verwaltung*, 3, pp. 95–102.

Wagner, W.E. (1995) 'The science charade in toxic risk regulation', *Columbia Law Review*, 95 (7), pp. 1613–1719.

Weber, S. (1991) 'Environmental information and the European Convention on Human Rights', *Human Rights Law Journal*, 5, pp. 177–185.

Winter, G. (1990) *Öffentlichkeit von Umweltinformationen. Europäische und nordamerikanische Rechte und Erfahrungen*, Baden-Baden: Nomos Verlagsgesellschaft.

Winter, G. (1994) 'Von der ökologischen Vorsorge zur ökonomischen Selbstbegrenzung', *Aus Politik und Zeitgeschichte*, 37, pp. 11–19.

Xiberta, J. (1994) 'The eco-management and audit scheme', *European Environmental Law Review*, 3(3), pp. 85–89.

5

DEREGULATION AS AN ENVIRONMENTAL POLICY INSTRUMENT IN HUNGARY

Gyula Bándi

INTRODUCTORY REMARKS

In its general meaning, the term deregulation means either to diminish the field of legal regulation, or to replace it with other means and methods. However, when examining the issue of deregulation and the environment in relation to the Central and Eastern European (CEE) countries (also frequently referred to as the countries in economic transition – CIT), it has to be kept in mind that these countries currently have a less detailed and less sophisticated environmental policy and legal framework than would be desirable. This is reflected in the assessment by the Director of the Regional Environmental Center for Central and Eastern Europe (REC), who in a report on the approximation of EU environmental legislation, wrote the following:

> After successfully negotiating Association Agreements with the EU, ten CEE governments have begun adjusting internal economic structures and developing mechanisms needed to proceed towards European integration. This complex process calls for a comprehensive strategy which may allow the prioritisation process needed to achieve the final goal. Among all the important areas of adjustment, environmental protection was identified as one of the most difficult and resource intensive.
>
> (REC, 1996, p. 7)

This is certainly true in the case of Hungary, which forms the focus of this chapter. Consequently, we have to understand first of all, if, when and under which circumstances it is possible to speak of deregulation in the environmental area. I take the approximation problems as a core phenomenon, because both in political and economic areas, accession to the EU is of overriding interest to the CEE countries. Hence, the chapter considers the approximation to EU legislation as a cornerstone problem.

Ludwig Krämer, a well-known expert in EU environmental law, summarises the regulatory problems deriving from this endeavour as follows:

> Experience shows that the adoption of the text of EC environmental legislation by a new acceding country is not the real problem. Rather, EC environmental law increasingly requires administrative procedures to be set up, clean-up and monitoring plans to be elaborated and progressively put into practice and authorisation procedures, monitoring, surveillance and control mechanisms to be instituted.
>
> (Krämer, 1996, pp. 435–436)

Administrative or authorisation procedures or control mechanisms all need specific regulations. Furthermore, it is quite clear that the 'adoption of the text of EC environmental legislation' may need new regulatory efforts. In the CEE countries, in a situation in which environmental issues do not represent a priority interest, there are two different attitudes of the state in relation to environmental regulation, which need to be made compatible. On the one hand, there is the recognition of a need for environmental regulation in order to develop a proper system of environmental law, enforcement and compliance. On the other hand, there is a preference that this should be done within the framework of a developing market economy, in which the utilisation of self-regulatory instruments and market incentives should be present. The second objective would certainly, at least partly, require deregulation in the longer run.

Before entering into the discussion of the Hungarian situation, it is useful to first take a wider perspective and have a look at the motivation of the CEE countries to join the EU. Qualifying for membership of the EU means that these countries have to develop their own environmental policy system, have to extend environmental regulations to neglected areas, as for example waste management, and have to apply new types of EU legislation, such as voluntary environmental management schemes. The REC paper states:

> One of the main driving forces behind the process of approximating CEE environmental policy and legislation to the respective policy and legislation of the European Union is the strong political commitment of the CEE countries.
>
> (REC, 1996, p. 16)

The author had the honour to be on the Editorial Board of the REC's approximation report, and the findings of the document are used in this chapter to present examples of the state of environmental law in the CEE countries, including Hungary.

CEE ENVIRONMENTAL PROVISIONS: GENERAL FINDINGS

This section provides a short overview of the most important findings of the above mentioned REC document, to outline the main types of regulatory provisions and tasks found in the CEE countries.

Constitutional-type provisions

Most of the constitutional provisions which have been adopted over the past four to six years in the region contain a direct reference to a clean environment as a basic human/citizen's right. In some cases, environmental protection is listed as one of the fundamental tasks of the state. For example, the Hungarian Constitution states in article 18: 'The Hungarian Republic recognises and enforces the right to a healthy environment for everyone.'

The provisions on the right to a clean environment are usually linked either to the protection of human life or health, or to state tasks. Neither of them can be interpreted in a way that would allow their use in legal procedures. Both types of provisions are of more theoretical than practical importance.

Furthermore, most of the CIT countries have their own general or framework environmental act. These acts are sometimes very general or even vague, but they can include one or several legal instruments or procedures. Two of the latest examples of this kind of legislation are the Slovenian Environmental Act of 1994 and the Hungarian Environmental Protection Act of 1995. This new generation of environmental acts seems to be a growing trend in the region and has to be evaluated as positive. As Kiss and Shelton (1993) have highlighted: 'The problem of fragmentation and conflicting laws is a current one and there is considerable need for a coherent, integrated approach to environmental codification' (Kiss and Shelton, 1993, p. 15).

The development of environmental acts is an important step towards a more coherent approach. There has also been a tendency to develop national environmental policies or programmes, similar to the Environmental Action Programmes of the EU. These policies or programmes either exist in a draft format only or they are very general, without direct reference to implementation. There is a trend towards more detailed action plans, but usually these are in a draft or abstract form only.

General environmental policy regulation

Due to the fact that most of the countries of the region have adopted new environmental laws in recent years, it is not surprising to find that the field of environmental policy is one of the most advanced. The easiest step in the process was the adoption of some general principles of environmental law.

Table 5.1 The inclusion of environmental policy principles

Principles	Czech Republic	Council of Europe	EU	Hungary	Slovenia
Precaution	—	Yes	Yes	Yes	—
Prevention	Yes	—	Yes	Yes	Yes
Substitution	—	Yes	—	—	—
Biodiversity	—	Yes	Partly	—	—
Sustainable development	Yes	—	Yes	—	—
Ecological stability	Yes	—	—	—	—
Acceptable load	Yes	—	—	—	—
Information	Yes	Yes	Yes	Yes	Yes
Public participation	—	Yes	—	—	—
Non-degradation of natural resources	—	Yes	—	—	—
Cooperation	—	Yes	International	Yes	Yes
Education, awareness raising	Yes	—	Yes	Yes	—
Liability, causality	—	Yes	Yes	Yes	Yes
Integrated protection	—	—	Yes	Yes	Yes
Restoration of damages	—	—	—	Yes	Yes
Compulsory insurance	—	—	—	—	Yes
Incentives	—	—	—	—	Yes

However, some of these principles, such as rectifying pollution at source or the precautionary principle, need further elaboration and putting into practice.

Table 5.1 compares the inclusion of various environmental policy principles in the national Environmental Acts of three CEE countries (Czech Republic – Environmental Act of 1991; Hungary – General Environmental Act of 1995; Slovenia – Environmental Act of 1994) with the Draft Model Act of Environmental Protection, drawn up by a group of experts under the Council of Europe heading in 1994, and the EU's environmental policy principles, as presented in the Environmental Action Programmes and in the Treaty provisions.

General environmental instruments

As far as general environmental instruments are concerned, the REC report considered three types of instruments: Environmental Impact Assessment (EIA), access to information, and environmental management systems. Legislation on EIA has developed rapidly in the CEE countries in recent years. However, the adoption of the legislation has not always been followed by the implementation of EIA procedures. This part of the legislation is still under development.

Access to information is very important as a precondition for public participation. Full free access is mostly absent, due to the fact that the former socialist legal systems did not integrate public rights into different areas of decision-making. However, some progress is being made, as later sections will demonstrate.

The development of environmental management schemes requires a well-established market economy. It would thus not be realistic to expect these countries to have an advanced form of voluntary environmental management. Bulgaria, Hungary, Romania and Slovenia are the only examples where one can witness the first examples of utilisation of these schemes. The Hungarian example will be discussed in more detail later.

Protection of air

The situation is satisfactory as far as the setting up of plans for the improvement of air quality is concerned. Point sources usually receive sufficient attention in the legislation and emission monitoring systems are in place. Generally, the use of BATNEEC as a technology requirement in air emission regulations is missing in the countries of the region, or one may find only some early drafts of proposals in this respect. The basis of air pollution regulations is to require an administrative authorisation procedure in the case of those industrial plants whose operation is likely to cause pollution. Most of the countries have similar authorisation procedures to those in the EU. As far as possible improvement policies and strategies relating to industrial plants are concerned, the situation is less satisfactory.

Chemicals, industrial risks and biotechnology

The use of chemicals, the chemical industry and other types of hazardous industrial activities have received substantial state aid for decades. Furthermore, these activities have been shrouded in secrecy. However, this is a field where immense changes can be achieved by simply adopting the EU's legislative provisions into domestic legal systems.

In the case of dangerous substances or chemicals, provisions are reasonable. Both packaging and labelling are regulated, and registration and permitting procedures are in place. Most of these rules are based on the regulation of toxic materials and public and worker health. Hazardous industrial activities have had much less attention to date, although there is some draft legislation on the control of these activities and on the need for developing emergency requirements. Generally, the public is not being informed about hazardous industrial activities and emergency response

79

plans. The lack of public information is closely linked to the missing access to environmental information in general.

Nature conservation

Nature conservation has the longest history of any field of environmental law. As nature conservation is relatively independent of economic development issues and the political conditions of a given country, this is a field of environmental law in which the CEE countries are most advanced when compared to EU Member States. The concept of safeguarding protected areas and identifying protected species are traditional areas of legislation. The methods differ from country to country, but existing structures of protection can be found everywhere.

Noise abatement

The relative low level of ranking of noise abatement issues on the approximation scale reflects a difference of legislative concepts. In the EU, noise abatement focuses on product standards. At the domestic level, noise abatement regulations focus on local impacts and the designation of noise abatement zones is the principal basis of legislation, while product standards are not that important. Another reason is the relative back-wardness of market economy and the lagging behind of product standards in the CEE countries.

Waste management

Waste management regulations represent another weak area of environ-mental law in the ten CEE countries. To some extent, this is a consequence of the lack of regulations in the fields of chemicals and hazardous industrial activities. A second reason is that the CEE countries generally do not apply such concepts as life-cycle analysis of products. A third reason is the relatively low value of primary raw materials, resulting in little interest in developing waste-minimising technologies or recovery and reuse options. Hence, the regulation of waste management has received little attention from legislators. This is exemplified by the lack of specific landfill or incineration requirements. The granting of permits is the best way of preventing the harmful effects of waste disposal. In most of the CEE countries some authorisation procedures exist, but generally limited to hazardous wastes.

Water

Water quality is one area for which regulation is more advanced. This can be attributed to the long history of legal issues related to water, dating back to the last century, the need for international cooperation, and the relative visibility of the problem. Water discharges are usually regulated. The regulation of water quality generally covers drinking water, but very rarely bathing water. The REC report did not compare the exact quality requirements.

Summary

To provide a general summary of the situation regarding environmental regulation in the CEE countries, the REC report lists the following achievements and problems (see Table 5.2). The problems have to be kept in mind when discussing the scope for deregulation. The next section will explore this issue in more detail, using the example of Hungary.

REGULATION AND DEREGULATION: HUNGARIAN EXAMPLES

In view of the findings on environmental regulation in the CEE countries, one might presume that any consideration of deregulation is limited by the massive regulatory urgency present in these countries. In the following sections, the discussion will focus on Hungary. In order to examine the problems of regulation versus deregulation, two specific areas are taken as examples:

- public participation;
- environmental management schemes.

In both of the cases, the examination centres around the provisions of the new Hungarian Environmental Act (Act LIII of 1995), which came into force at the end of 1995. The reasons for selecting the two cases studies are as follows. In the case of public participation, the issue is closely linked to the democratisation process as a whole and to issues such as openness of

Table 5.2 CEE achievements and problems

Achievements	Problems
• Political commitment	• Unbalanced development of law
• Coordination on the state level	• Lack of a clear approximation strategy
• Progress in selected areas	• Problems with access to the EU
• Network of professionals	legislative process
• Business interests	• Incomplete law drafting and frequent
• Translation of EU environmental Directives into local languages	changes of law
	• High costs of approximation
	• Poor enforcement

Source: REC, 1996, pp. 16–18

government and public administration. These issues are also relevant in the deregulation context, as the chapter by Heyvaert has shown.

In the case of the second topic, there are a number of arguments that support the selection. Under the conditions of a market economy, an increasing emphasis has to be put on market forces and the 'self-regulation' of the economy. Consequently, market-oriented measures in environmental policy should be strengthened. Environmental management schemes are a main instrument of self-regulation. Thus the two different subjects reflect different elements of the political, legal and economic transition, as relevant to environmental protection.

Public participation

Public participation is a mixture of three elements, as highlighted by the UN Economic Commission for Europe (1996) guidelines on public participation:

- access to environmental information;
- participation in rule and decision-making processes;
- access to justice.

When we examine these elements in more detail, we find that the problems encountered in establishing public participation processes are rooted in the difficulties of the past fifty or so years. As already mentioned earlier, the general attitude of the socialist state was to neglect public participation. This was justified through the claim that socialist governments have as a primary purpose the satisfaction of the public interest, hence there was no need to make specific provisions for public participation. The importance of public participation is well characterised by the following quote:

> No legal system presently appears to be able to offer truly satisfactory solutions to this challenge . . . Flexibility and pragmatism lead to a degree of uncertainty and leave many regulatory gaps. Rigidity and over-regulation do not succeed in eliminating the gaps and are apt to lead to enforcement problems that make them inefficient.
>
> (Lambrechts, 1996, p. 107)

Providing for public participation is not the easiest task the CEE countries face. If we take the Hungarian legal system as an example, administrative procedures are far from transparent, information is still restricted to the parties closest to the case and there are no procedures for requesting a public hearing.

A number of steps have already been taken to open up the system. For example, access to the courts has been widened. New legislative acts have been passed concerning the protection of personal data and access to data of public interest. In the field of environmental law, public hearing procedures have been introduced within the context of the EIA regulations. The 1995

Environmental Act provides for 'actio popularis' (public action) in the fields of administrative procedure and civil procedure. Furthermore, there are provisions allowing public involvement in environmental law-making.

There is a separate chapter in the 1995 Environmental Protection Act entitled 'Citizen Participation' (Chapter VIII). The main provisions in the chapter were tailored to the role of environmental associations. These associations are given the right of a party in administrative proceedings. Besides participation in decision-making, there are two more essential 'actio popularis'-type procedural rights given to these associations:

- the right to request government agencies or local governments to take appropriate measures which fall within their scope of authority;
- the right to file a lawsuit against a 'user' of the environment, either in order to prohibit unlawful conduct, or to issue an obligation to take the necessary measures to mitigate damage.

Public participation requires a detailed set of guarantees provided by regulation. In connection with the participation of the public in decision-making, this includes the need to regulate:

- the details of how the information on decision-making is transmitted to the public (for example, the sphere of the interested public should be defined, the methods of offering information – mail, media, local news-board, etc. – at least, should be listed, the proper time-frame open for the public to study the information should be fixed, etc.);
- the scope or extent to which the public should have access to the decision-making process itself;
- the methods of public involvement – as an observer, as a party, as a plaintiff, etc.;
- the way in which the public is to be informed about the results of their participation;
- the means and methods of legal remedy, in case there are some procedural infringements or disagreements;
- the role of different organisations and associations.

As all the above issues would guarantee the proper implementation of the right and duty of public participation, these should either already be a part of the given legal system or should be regulated. This is particularly important in countries like Hungary, where neither the pre-war, nor the post-war system took care sufficiently of public rights.

The REC also has carried out a major project on promoting institutions for public participation. In the project report, the authors noted that in the countries of the region 'the number of legal drafts and amendments is so great that because of this the transition, at least from a legal point of view, would be a long lasting process in these countries' (Stec and Tóth Nagy, 1994, p. 14).

In practical terms, there are opportunities for using public participation procedures as a good foundation for future deregulation moves. Public participation has to be regarded not as an end in itself, but as mainly a kind of basic procedural instrument. It has several positive aspects which could support deregulation:

- the proper utilisation of public participation means and methods may give the public authorities a better image;
- there might be fewer tensions between authorities and the public as a result of participation;
- through a discussion with the public, the developing environmental legislation may be better understood, thus facilitating implementation;
- the public may be a good source of information, at least as far as local interests are concerned. Additionally, NGOs can be good providers of scientific information;
- there may be scope for the public assisting in enforcement (watchdog functions).

The requirements for very detailed regulations on public participation could thus easily become the basis for deregulation. While in Hungary detailed regulations are needed, in those countries where public participation is already in existence, the regulation requirements are not so important. We should also not forget that the concept of public participation requires a new way of thinking within the public administration. As Weale (1992) has pointed out 'the effect of these trends is to question the operating principles of the traditional environmental policy process'.

Environmental management

A new trend in environmental protection is to rely increasingly on the self-regulation of business, which, as Ryding (1992) argues, will coincide with long-term business interests.

> In business terms, this is a process equivalent to generating the maximum income from a given stock of assets without depleting the capital base. From the perspective of business, it is important to fully integrate the value of environmental assets into operations . . . to guarantee conservation of nature for future generations.
>
> (Ryding, 1992, p. 525)

If environmental regulation is well developed and enforced, there are better chances for the development of self-regulatory elements, letting businesses create their own methods of environmental protection. If we add the international business dimension, including the search for new markets and the benefits business gets from European integration, businesses might be prepared to accept environmental regulations.

In business terms, Hungary has now privatised over 50 per cent of the previously state-owned enterprises. However, state ownership is still substantial. Additionally, much of privatisation has focused on small industries and resulted in employee buyouts. Both state-owned and small-scale businesses have difficulties in developing self-regulatory processes. Ryding (1992) emphasises that:

Developments in the business sector are influenced by various guiding inputs, including:

- laws, rules and regulations;
- public opinion;
- capital market;
- labour market and employee attitudes towards the corporation;
- consumers;
- dissemination and information (including advertising, government advisories, etc.);
- competition.

These guiding inputs are not always compatible, and can give conflicting signals. It is of vital importance to harmonise these signals with the goals of sustainable development. However, adopting a market-based approach will be a difficult task, because some business sectors will be made redundant by charging the full price (including environmental concern) of goods and services . . .

Regulatory instruments should be created to produce general economic conditions which enable a fair competition between companies that do not inhibit ecological progress from being made.

(Ryding, 1992, pp. 525–526)

One regulatory instrument that has been developed in the EU is Council Regulation 1836/93 on Environmental Management and Audit Schemes (the EMAS regulation). In Article 3, the regulation gives a full and detailed list of the environmental management scheme. The companies must be registered as a unit which may participate in the Community eco-management scheme. The competent authorities shall register the given company on the basis of a validated environmental statement.

In order for a site to be registered in the scheme the company must:

- adopt a company environmental policy,
- conduct an environmental review,
- introduce an environmental programme,
- carry out environmental audits,
- set objectives at the highest appropriate management level,
- prepare an environmental statement,

- have their environmental policy, programme, management system, review or audit procedure and environmental statement or statements examined to verify that they meet the relevant requirements of these Regulations,
- forward the validated environmental statement to the competent body of the Member State where the site is located.

One of the main elements and the starting point for the development of an environmental management system is environmental auditing. This has at least two main purposes and may require a general set of underlying regulations. First, auditing provides a background for setting up the proper environmental management structure in a company. Second, auditing may also be taken as a source of information for record-keeping within the company and reporting to the environmental authorities.

In order to promote the self-regulatory aspects of environmental auditing, the new Hungarian Environmental Act (Articles 73–76 and 78–81) introduced an obligation of environmental supervision or review, as a new element of the environmental tasks of public authorities. The requirement for environmental supervision or review is in effect an obligatory environmental audit requirement. The reviews 'must be carried out to disclose the environmental impact of certain activities and check the compliance with environmental protection requirements' (Article 73 Par. 1).

The review is carried out for ongoing activities, which in the case of new projects would have been subject to an environmental impact assessment procedure, or if these activities are proven or likely to be dangerous to the environment in highly protected areas, protected areas or protective zones. This review shall be ordered by the environmental authority (Article 74 Par. 2). The exact circumstances of this order have not been specified to date.

The review resembles an environmental impact assessment procedure. However, as it is dealing with ongoing activities, the environmental impacts must be demonstrated with exact data, and there might be no alternatives to these activities. The review should also describe measures taken or planned for the prevention or mitigation of environmental damage. The following decisions may be taken by the environmental authority on the basis of the review (Article 79 Par. 1):

- authorisation to pursue the activity (permit for operation),
- in conjunction with the permit, obligations for certain environmental protection activities,
- restriction, suspension or prohibition of the activity.

If a company wants to have official approval of its auditing procedures, the 1995 Act also makes this possible. In this case, the company may voluntarily undertake the environmental review process – or at least an audit similar to that process – and ask the environmental authority for their approval. Companies may want to opt to do this when they are in the process of being privatised.

Another regulation of an environmental auditing type became law after the 1995 Environmental Act. In implementing the very vague provision of Article 48, paragraph 4 of Act No. IL of 1991 on bankruptcy, liquidation and final settlement processes, the government adopted through Decree 106/1995 (IX. 8.) Korm. some special environmental and nature protection requirements.

There are two main options for undertaking an environmental review in a company under the bankruptcy or liquidation proceedings. Firstly, the manager of the company can send an environmental statement to the environmental inspectorate. Secondly, an environmental audit can be carried out. With this procedure, the first step is again to provide the environmental inspectorate with a statement. It may then decide to oblige the company under liquidation to carry out an environmental audit. The inspectorate, after consulting with some other authorities, such as the public health service or the water management authorities, may require the company to do so in the following cases:

- if the company otherwise should have been ordered to carry out an EIA;
- if, on the basis of the environmental statement, one can assume a significant environmental damage or environmental danger;
- if the inspectorate or other authorities have significantly different information on the company's environmental impacts than the information provided in the statement.

On the basis of the statement or the audit, the environmental inspectorate may order the company to settle environmental damages. The liquidator is obliged to enforce these orders. This may be done by contracting out the necessary remedial activities, or making an agreement with the purchaser of the company to attach these tasks to the deal.

All the above mentioned provisions with an environmental audit character form part of a wider system of environmental management. The 1995 Hungarian Environmental Protection Act also contains some reference to environmental management itself, requiring some companies to employ an environmental manager or official. The list of those companies that have this obligation is under preparation. However, the employment of one specialist will not change a company's environmental management system immediately.

Thus, we are seeing regulatory trends similar to those in the public participation case. The setting up of environmental auditing and management schemes begins with legal regulation and legal obligations. Subsequently, it is hoped that the employment of an environmental 'agent' or the utilisation of the obligatory or voluntary environmental review procedures will catalyse the development of further environmental management systems. Once environmental management is in place, the implementation of environmental requirements will be better.

REGULATION AND DEREGULATION AS ENVIRONMENTAL POLICY TOOLS

In a recent paper, Pieter Glasbergen dealt with the problems of the system of regulatory control. He wrote as a conclusion:

> we argued that administrative reform is imperative if we are to develop a more effective approach to environmental issues. We defined administrative reform as an issue of control . . . The operation of the classic model of control depends upon the input of a large and expensive legislative, executive and enforcement apparatus. In addition, this model puts some restriction on the capacity of control. A review of the characteristics of the control situation led us into a discussion of network management. This is a form of interactive control that seems to offer opportunities to mobilise special interests to support environmental reform.

Later the same author says:

> Network management will not entirely replace the application of the regulatory model of control. Interactive forms of control either precede the classic form of control or supplement it once it is in place.
>
> (Glasbergen, 1996, pp. 198–199)

Thus new challenges to environmental control require a solid regulatory background. Consequently, we have to emphasise that both regulation and deregulation are tools or methods of environmental policy and not the instruments of it. Hence, within the context of this paper, the term 'environmental policy instrument' is understood in a broad way. The real policy instrument question would be 'what' should be regulated or 'why' should we deregulate? Like public participation, deregulation should not be considered an end in itself.

In the field of environmental policy, the first task is to set up a proper regulatory framework which guarantees, or at least makes it possible, that the polluters prevent or reduce pollution, change their polluting activities, avoid or mitigate damage or restore damage to the environment.

In EU environmental policy, the Fifth Environmental Action Programme professes more support for the market-oriented and self-regulatory aspects of environmental policy than for the command-and-control type of regulation. This represents a shift in the perception of the role of the state, from the major regulator to the primary guardian of environmental interests.

In the following section, we try to present our view of the regulation–deregulation debate in two simple models, in which we use the term 'command-and-control regulation' for the classic model of state administrative influence; 'market-oriented instruments' for the use of economic and financial incentives or disincentives; and the term 'self-regulation and eco-management' for those instruments such as the EU EMAS regulation.

We may illustrate the shift from command-and-control model to deregulation in a very simplified way as follows:

| Command-and-control regulation | ⇨ | Market-oriented instruments | ⇨ | Self-regulation or eco-management | ⇨ | Deregulation |

The time and effort spent on each of the first three elements is an important issue. In countries like Hungary, most of the first step is still missing, while we have to work on the second and the third step. However, the steps do not have to be taken in sequence but can be taken together. This is a necessity in the case of Hungary, so as to set up an efficient regulatory framework and to stimulate the market and business players at the same time. Consequently, our simplified picture would look very much like the following:

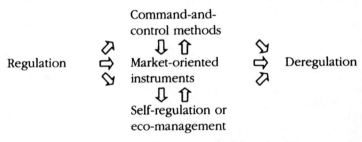

We should not forget that while in several fields of environmental policy and law, deregulation may be taken as a short-term policy instrument, in other fields of environmental policy and law, regulation must come first. Furthermore, there are certain fields of law, for which we cannot really consider deregulation, such as the field of liability. Finally, there are certain situations where regulations have to be quite extensive.

We do not want to suggest that due to the relative backwardness of environmental policy and legislation in the CEE countries, we have to regulate first in order to have something to deregulate in the future. Instead, there may be an opportunity to combine the two, or in other words, to regulate only to a level which is absolutely necessary. In some fields, there is a chance to use self-regulatory tools such as environmental management methods.

Theoretically, deregulation could mean controlled or limited regulation. However, in view of the current shortcomings in Hungarian environmental policy and law, there is not much scope for proceeding in this direction. The controlled or limited type of regulation would need a carefully planned regulatory strategy that is generally missing in Hungarian legislation. During the next few years, the task of restructuring at least partially the existing legal system and that of developing a new system, have to be carried out simultaneously. At the same time, we are also seeing legislation that is highly influenced by short-term political or economic interests. Thus the future of deregulation as an alternative to regulation is far from clear.

REFERENCES

Economic Commission for Europe (1996) *Guidelines for Public Participation in Environmental Decision-Making*, Geneva: United Nations.

Glasbergen, P. (1996) 'From regulatory control to network management', in Winter, G. (ed.), op. cit.

Kiss, A. and Shelton, D. (1993) *Manual of European Environmental Law*, Cambridge: Grotius Publications Limited.

Krämer, L. (1996) 'Adaptation to EC environmental rules: options for Central and Eastern European states', in Winter, G. (ed.), op. cit.

Lambrechts, C. (1996) 'Public participation in environmental decisions', in Winter, G. (ed.), op. cit.

REC (1996) *Approximation of European Union Environmental Legislation – Case Studies of Bulgaria, Czech Republic, Estonia, Hungary, Latvia, Lithuania, Poland, Romania, Slovak Republic and Slovenia*, Budapest: Regional Environmental Center for Central and Eastern Europe.

Ryding, S.-O. (1992) *Environmental Management Handbook*, Amsterdam/Oxford: IOS Press/Lewis Publishers.

Stec, S. and Tóth Nagy, M. (1994) 'Public participation in Central and Eastern Europe', in REC (ed.) *Manual on Citizen Participation in Environmental Decisionmaking*, Budapest: Regional Environmental Center.

Weale, A. (1992) *The New Politics of Pollution*, Manchester: Manchester University Press.

Winter, G. (ed.) (1996) *European Environmental Law – A Comparative Perspective*, Aldershot: Dartmouth.

Part II

SECTORAL PERSPECTIVES

6

LIBERALISATION IN THE ENERGY SECTOR

Environmental threat or opportunity?

Ute Collier

INTRODUCTION

While government intervention in the energy sector has traditionally been strong, both through public ownership and regulation, recent years have seen some fundamental changes in a number of countries, focusing in particular on privatisation and liberalisation. On the one hand, the interest in liberalisation has formed part of a general move to reduce state involvement in industry (especially the utility industries), through privatisation and/or by opening markets to competitors. On the other hand, liberalisation has been an essential ingredient of the completion of the internal market in the European Union (EU). Energy has been considered an important commodity for which free movement should be achieved. In both cases, deregulation is an important component of the strategies, although in reality, the issue is effectively one of reregulation rather than deregulation as such.

The rationale for liberalisation has been based primarily on economic arguments, concerned with reducing market barriers and increasing economic efficiency. Environmental issues have rarely been considered relevant to this discussion. With the significant role of energy production and use as a causal factor in many environmental problems (in particular climate change and acid rain), the environmental ramifications of a liberalised energy market are an important issue for policy analysis.

Liberalisation potentially has both positive and negative environmental implications. On the one hand, it promises an opportunity for shaking up a monopolistic energy market, dominated by supply-oriented companies, focused on large-scale technologies. This could provide a chance for renewables and demand-side management (DSM), two options with substantial environmental benefits. On the other hand, a 'free for all' situation could mean damaging competition between options such as gas

93

heating versus district heating or gas-fired power stations versus renewables. The best outcome in environmental terms cannot be assumed, especially while multiple market failures exist, including those related to external costs.

The aim of this chapter is to assess the environmental implications of energy sector liberalisation, with a particular focus on the electricity sector, on account of its crucial role in emission abatement. It commences with a summary of the state of energy sector liberalisation in EU Member States. It then reviews the theoretical issues related to liberalisation and the environment, before specifically looking at two contrasting cases: the privatised, liberalised electricity sector in the UK and the monopolistic, local energy companies in Germany.[1] The chapter argues that there is an urgent need to ensure that environmental considerations are not ignored in the liberalisation process. Liberalisation can offer opportunities for emission reductions but, to maximise them, an appropriate regulatory framework is crucial. At the same time, under certain circumstances, liberalisation can be potentially damaging and should not be pursued at all costs.

LIBERALISATION AND THE ENERGY SECTOR IN THE EU

In the EU, as elsewhere, the energy sector has traditionally seen heavy government intervention, both through public ownership and regulatory measures. Justifications for this heavy intervention include the essential nature of energy supply in the economy and the natural monopoly characteristics of the energy sector. However, in recent years, the energy sector has been undergoing a period of dramatic change worldwide. In the EU, a number of Member States are reforming their industries or evaluating their structure, ownership and regulation.[2] Some are considering whether to liberalise, and if so to what extent. Furthermore, in June 1996, proposals for electricity sector liberalisation within the framework of an Internal Energy Market were agreed at EU level, leading to a gradual opening up of Member States' energy markets to competition.

Structure, ownership and regulatory frameworks of the energy sector have always varied between Member States and these differences have been a major obstacle to the completion of the Internal Energy Market. Recent changes have also taken different forms in different Member States. Out of the EU Member States, the most radical change has taken place in the UK, where the energy sector has now almost completely been privatised, as well as partially liberalised. At the other end of the spectrum, the monopoly power of Eléctricité de France is likely to remain more or less intact, especially in view of France's commitment to nuclear power.

In Italy, efforts are currently underway to privatise the state-owned energy monopolies. ENEL, the electricity monopoly, was transformed into a joint stock company in 1992. Plans for its eventual privatisation have been amended several times and it is currently not clear to which extent the market will be liberalised. A law on the establishment of an electricity regulator was adopted in 1995, which requires the regulator to balance the interests of companies with those of environmental protection, resource conservation and the use of renewable energy sources. It remains to be seen how seriously these objectives will be pursued. In Sweden, plans for liberalisation were put on halt by a change of government. Meanwhile in Spain, partial privatisation over recent years has not changed the monopoly position of individual firms and there are no plans for greater liberalisation (for more detail on these countries see Collier and Löfstedt, 1997).

In Germany, there are both regional and local monopolies, protected by demarcation contracts. However, proposals for liberalisation are currently under discussion in the German parliament and are likely to benefit the eight large regional companies[3] (run as limited companies, partially privately owned with some share ownership by the federal states), at the expense of local, municipally owned companies. As will be further explored later, this could have detrimental effects on environmental performance. Local (and sometimes regional) authorities also play an important role in energy policy and own energy companies in Austria, Denmark and Sweden.

ENERGY SECTOR LIBERALISATION AND THE ENVIRONMENT – SOME THEORETICAL ISSUES

The main component of liberalisation is the introduction of competition in a market previously dominated by a monopoly or a very limited number of companies. In the energy sector, these monopolies have generally been considered 'natural', meaning that because of the nature of the market (e.g. high investment costs, economies of scale), no competition is possible. Natural monopolies essentially apply to the transmission and distribution side, as not more than one power grid and one gas network makes economic sense. However, even here competition can be created by enforcing access to the grid through regulation. Also, there can be competition between fuels, especially in the heating market where fuel substitution between different options (oil, gas, district heating and electricity) is possible. On the electricity generation side, a kind of natural monopoly used to exist when large-scale fossil fuel or nuclear power stations were the most viable generating option (although often only because of large government subsidies), due to the huge sums of investment required. However, slower growth rates in electricity consumption, the potential for demand-side management (DSM), as well as the attractiveness of combined-cycle gas turbines (CCGTs), small-scale combined heat and power (CHP) plants and some small-scale renewable energy sources mean that natural monopoly arguments no longer apply (see Jaccard, 1995).

The effects of competition

Liberalisation in the energy sector is concerned with introducing competition into at least part of the market. Proponents of competition assert that it creates an incentive mechanism for more efficient operation, as well as opportunities for innovation (OECD, 1994). As Arentsen and Künneke (1996) point out, supporters of liberalisation are concerned with increasing *economic* welfare at the *macroeconomic* level. It is thus not entirely surprising that there are various microeconomic, social[4] and environmental side-effects, both positive and negative.

Energy sector liberalisation can take different forms, amongst the most common of which are:

- some competitive procurement on the generation side by a vertically integrated monopoly energy company (e.g. under US PURPA);
- vertical disintegration and full competition on the generation side;
- vertical disintegration and full competition in both generation and supply.

The UK model is aiming to achieve the latter, while many other countries are moving towards at least some form of competition on the generation side. Within these systems, ownership and governance can vary considerably, as Arentsen and Künneke (1996) examine in more detail. Despite the variety of systems, some general observation about effects on technology and fuel choice can be made. According to the OECD (1994) these include the following:

- less investment in large generating plant;
- increased diversity of investors, technology and fuels;
- competition between fuels in the heating market;
- increased development of smaller units, typically closer to load centres.

The OECD study found that investment decisions in a liberalised energy market will tend towards technologies that are less capital-intensive and that have shorter lead times. Furthermore, increasing the number of investors can also encourage the consideration of less traditional, financially riskier, technologies. Currently, competition favours CCGTs, the economic attractiveness of which depends mainly on perceptions about gas prices and availability.

The potential environmental consequences, as well as the general welfare effects, of liberalising energy markets are complex and, to date, there has been little analysis of these issues. In environmental terms, the implications of liberalisation in the following areas are crucial:

- choice of generation technology;
- attractiveness of the promotion of energy efficiency.

Monopoly utilities (especially large ones such as Eléctricité de France) have generally focused their investments on large coal-fired and nuclear plant, and have shown little interest in small-scale, environmentally friendly generation options based on renewable energies or combined heat and power (CHP) plants. Because of their small scale, these technologies are particularly attractive for auto-generators but access to the grid to sell any surplus electricity has been difficult in many countries. Hence, the liberalisation of generation and access to the grid is often regarded as positive in environmental terms (see e.g. Flavin and Lenssen, 1994), as it tends to result in investment in plants which, although by coincidence rather than design, have environmental benefits. Furthermore, competition can also be beneficial for demand-side management (DSM), although the US experience has shown that there is a need for specific regulatory provisions to create a level playing field (see e.g. Nadel and Geller, 1996; Hirst, Cavanagh and Miller, 1996; Kozloff and Dower, 1993; Collier, 1994b).

However, competition can also have negative effects. Competition in the heating market, for example, does little to enhance the attractiveness of CHP, which in environmental terms is generally preferable to power-only plants. As Collier (1994a) and Rüdig (1986) have shown, CHP combined with district heating (DH) has been successful in Germany and Denmark, where competition has been specifically excluded, and municipal energy companies have been able to designate specific areas for district heating and/or have offered incentives for connection to the grid. This issue will be further discussed in later sections.

Another problem with liberalisation is the implicit aim to reduce energy prices (as a means of increasing the competitiveness of European industry). At the same time, low energy prices have been widely recognised as a main obstacle to greater energy efficiency (see e.g. European Commission, 1995). Hence, unless other incentives can be created, liberalisation is likely to have a negative impact from this point of view. At EU level, it has always been argued that any negative effects of the IEM could be solved through complementary environmental measures, in particular the carbon/energy tax (Collier, 1994b). However, in its absence, there is no real cost internalisation. As Collier and Löfstedt (1997) have shown, energy efficiency programmes are inadequate in a number of countries and large emission reduction potentials remain. Any further weakening of incentives for energy efficiency as a by-product of liberalisation would thus have to be deplored.

The relevance of public ownership

A further issue of potential consequence in a liberalised market is that of the priorities set by private companies, as opposed to those pursued by public energy companies. In most EU countries, at least part of the energy sector is currently in public ownership, either at national, regional or municipal level.

While in the UK, liberalisation has been accompanied by privatisation, this does not necessarily have to be the case. However, liberalisation would inevitably lead to an erosion of the dominant role of public energy companies, as private companies take up investment opportunities. The environmental implications of this depends on the current performance of public utilities.

Politically, as Majone (1994) discusses, public ownership is supposed to give the state the power to impose a planned structure on the economy and to protect the public interest against powerful private interests. Public interest theory suggests that public ownership could maximise net benefits to society, which in principle should include environmental protection. With private ownership, firms maximise profits by minimising costs in relation to output. Many environmental costs are external to this equation, thus leading to a sub-optimal resource allocation in environmental terms. If environmental concerns are a political priority, public enterprises can be used as a vehicle for realising policy objectives, including the internalisation of environmental costs.

However, Bishop, Kay and Mayer (1995) argue that public ownership has been largely ineffective as a means of reducing market failures and Majone (1994) has found that public ownership neither guarantees public control of nationalised enterprises nor necessarily provides levers and instruments of economic management. Examples such as performance of the now defunct UK Central Electricity Generating Board (CEGB) show that in environmental terms, public ownership has not necessarily had any benefit. In fact, in the case of acid rain this publicly owned company has actually had a negative influence on government policy (see Boehmer-Christiansen and Skea, 1991). A specific problem in many cases is that public energy companies are generally under the control of industry or economics ministries, which have not usually attached much priority to environmental concerns. Instead, government control has focused on issues such as security of supply, consumer protection and the safeguarding of indigenous fuels, especially coal (Arentsen and Künneke, 1996).

However, evidence of positive instances of public ownership also exists. As Löfstedt (1997) has shown, the Swedish state-owned utility Vattenfall implemented an ambitious energy efficiency programme (Uppdrag 2000) which was clearly politically driven. Now as Vattenfall is being forced to operate on a commercial basis, such programmes are no longer possible. Other positive examples can be found in the operation of municipal energy companies in a number of Member States (e.g. Austria, Germany, Sweden). In these cases, political control has positive effects due to the environmental priorities set by local politicians. The fact that these companies operate in monopoly situations and are not threatened by competition is important.

As an alternative to direct control through ownership, policy objectives such as environmental protection can be imposed on private companies through regulation. However, Flavin and Lenssen (1994) argue that while it is relatively straightforward to get a public utility to change direction, changing the regulatory formulae of private utilities is a much more complicated process. Furthermore, the problem of 'agency capture' can occur. While Flavin and Lenssen have a point, the crucial issue is that governments and regulators are becoming ever more interested in deregulation and competition, as their main objective for the energy sector. Other regulatory objectives are increasingly marginalised. Firms in this competitive system are ever more preoccupied with cost cutting and the maintenance of their market share. As Sioshani (1996) has shown, this is having a negative impact on DSM in the US. In Europe, where DSM is not well established, competition may well mean that it never will take off.

On balance, the evidence is that ownership can be a variable in the realisation of environmental objectives. However, as the following examination of the UK case shows, it is not so much the ownership issue per se, but rather the notion of a liberalised energy market which matters most in environmental terms. The rest of the paper will consider in more detail what effects liberalisation can have (the UK example) and how it might interfere with environmentally focused energy activities (the case of municipal energy companies in Germany).

LIBERALISATION IN THE UK

The organisational and regulatory framework

During the 1980s, the conservative government in the UK embarked on a large-scale programme of privatising public utilities, including energy companies. British Gas was privatised in 1986, while the electricity companies were sold in 1990/91.[5] The privatisation process was primarily ideologically driven and only during the later stages paid attention to issues of liberalisation. It almost entirely ignored the environment as a factor.

The UK's 'dirty man' image of the 1980s was directly related to the electricity sector, as the dominance of large coal-fired power plants (partially as a result of government policies supporting the coal industry), combined with little investment in energy efficiency and renewables, contributed to the UK being one of the largest emitters of SO_2, NO_x and CO_2 in Europe. As Boehmer-Christiansen and Skea (1991) have shown, the Central Electricity Generating Board (CEGB)'s negative stance on acid emissions was instrumental in shaping the government's approach to the problem. There was much potential for the regulatory framework of the privatised industry to include provisions for a better environmental performance. For this, the government only had to look at some of the well-documented examples from the US which show how regulatory authorities can coerce private companies through both regulation and incentives into environmentally beneficial activities (see e.g. Nadel and Geller, 1996; Hirst, Cavanagh and Miller, 1996; Kozloff and Dower, 1993; Collier, 1994b). Subsequent sections will reveal how this opportunity has essentially been missed.

Privatisation has resulted in a total upheaval of the energy sector which still continues seven years after the first companies were privatised.[6] Competition has had the biggest impact on the electricity generation side with the two largest of the CEGB's successor companies, National Power and PowerGen, having to concede an increasing market share to other companies. While independent generators have benefited to some extent, the biggest winners have been the distribution companies (Regional Electricity Companies – RECs, previously the area boards), who have heavily invested in their own gas-fired plants. Competition is also being introduced in the gas sector and by 1998, full competition will apply to the electricity distribution side, although it is not clear whether this will result in any great changes on the domestic consumer side. Finally, takeover bids by National Power, PowerGen and Scottish Power for some of the RECs, as well as potential mergers, are resulting in greater concentration and some vertical integration.

As in the case of other privatised industries, the government established quasi-autonomous regulatory bodies to oversee the industry. OFGAS is in charge of the gas sector, while the Office for Electricity Regulation (OFFER) deals with the electricity companies. The latter is directed by the government appointed Director General of Electricity Supply (DGES), the current incumbent being Professor Stephen Little-child, one of the original privatisation advisors to the government. As electricity privatisation coincided with the growing concern about environmental issues, culminating in the publication of a White Paper on the environment in 1990, a consideration of these issues in the privatisation legislation was to be expected. The Electricity Act does indeed contain some reference to environmental protection. The DGES has the duty 'to take into account the effect on the physical environment of activities connected with the generation, transmission or supply of electricity' (Electricity Act, 1989).

Furthermore, electricity generators and suppliers must have regard to 'the desirability of preserving natural beauty, flora and fauna' and must provide a statement on how they propose to achieve this. If they fail to do so, the regulator can intervene. These provisions are rather vague but could be used to justify more extensive environmental activities, should the need arise. Additionally, the Act included some provisions for energy efficiency and for the support of nuclear power and renewables. The next sections will aim to establish whether these provisions have proved adequate to date.

The dash for gas

There is no doubt that the most significant development in emission terms has been the so-called 'dash for gas'. Before privatisation, British Coal had a guaranteed market for a large part of its output and the CEGB had plans for a further expansion in coal-fired capacity. Since privatisation, CCGTs have become attractive as, while gas prices have been relatively low, they offer independent generators the most economic method of entry into the system. Furthermore, they present the established generators with a cost-effective way of meeting the obligations for reducing SO_2 and NO_x emissions under the Large Combustion Plant Directive, avoiding the expensive retro-fitting of coal-fired plants with flue-gas desulphurisation units. Between 1990 and 1995, CCGT capacity increased over one-hundredfold from 76 MW to 8540 MW (Department of Trade and Industry, 1995a). As a result, fuel use in electricity consumption is changing fast, as can be seen in Figure 6.1.

CCGTs have some clear environmental advantages. While they produce no SO_2 emissions, NO_x emissions are only about 0.10 grammes per kWh, compared to 1.29 g/kWh for an ordinary coal plant (Flavin and Lenssen, 1995). CO_2 emissions are less than half those of a coal plant, due to higher efficiencies and the lower carbon content of natural gas. These advantages are clearly reflected in the trends of these emissions since the first CCGTs came on stream in 1991,[7] as demonstrated in Figure 6.2.

The pronounced downward trend in these emissions since 1990 is almost entirely due to the decrease in coal burn, although in the case of SO_2 and

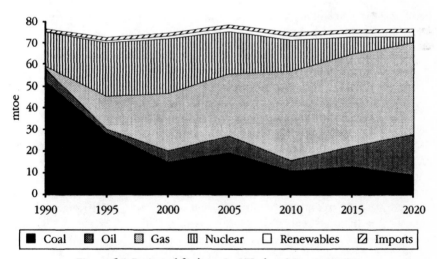

Figure 6.1 Projected fuel use in UK electricity generation
Source: Department of Trade and Industry, 1995b

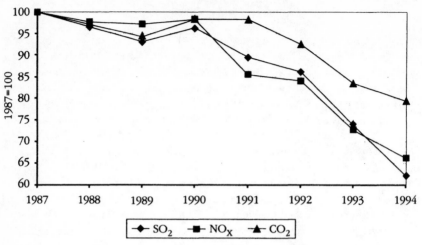

Figure 6.2 Emission trends from UK power stations
Source: Department of the Environment, 1996

NO_x, other factors such as the use of lower-sulphur coal and the introduction of pollution control equipment (at Drax and Ratcliffe power stations) also plays a role.[8] As further CCGT plant will come on stream until 2000, albeit at a slower rate, the downward trend in emissions will continue. This is particularly significant for UK CO_2 emissions, which despite emissions growth in the transport sector, are expected to fall by between 4.4 and 7 per cent by 2000 (Department of Trade and Industry, 1995b). Some of the investment in gas-fired plant undoubtedly would have taken place even without privatisation, although it is unlikely that it would have happened at the same pace and scale (Surrey, 1996).

While there will be substantial emission reductions, in particular the residual CO_2 emissions will be far from negligible. Notwithstanding the switch to CCGTs, the contribution of power stations to total CO_2 emissions is still expected to be around 24 per cent by 2000, compared to 34 per cent in 1990 (Department of Trade and Industry, 1995b). It can be argued that at least in the short-term, a switch to essentially CO_2 free sources such as renewables or nuclear power was not possible, so that a move to CCGTs was the best solution. However, even greater CO_2 reductions could have been technically possible by designing and operating CCGT plants as combined heat and power (CHP) plants or to invest more in small-scale CHP plant (gas or coal-fired). While CCGTs are much more efficient than coal-fired plant, their efficiencies still hover around 50 per cent. Hence, 50 per cent of the resource used continues to be wasted in the form of heat, while with CHP net efficiencies of up to 85 per cent can be reached and emissions reduced correspondingly. In

Germany, several new CCGT plants operate in CHP mode with high net efficiencies.

The main problem in the UK is the lack of a market for the heat. While in Germany and some other countries, government subsidies have been available for the construction of district heating grids, this has not been the case in the UK. Furthermore, district heating cannot compete against gas, especially in areas where gas networks exist already. However, there is still a good potential for smaller-scale CHP plants whose heat load can be used either in industry, the public sector or new housing estates. In the public sector, the budget constraints imposed on local authorities hinder greater investment in this area (Collier and Löfstedt, 1996).

Small-scale CHP investment has received some boost with liberalisation, although this has been totally dwarfed by the scale of the CCGT expansion. Between 1991 and 1995, CHP capacity increased from 2312 MWe to 3487 MWe (Department of Trade and Industry, 1996). The viability of CHP schemes, especially non-industrial ones, often depends on signing up potential customers. However, the forthcoming liberalisation of the supply market means that few customers are willing to sign long-term contracts in view of the fact that another supplier might offer electricity cheaper in the future. Competition on the supply side can thus act as an obstacle to CHP, while competition on the generation side can be beneficial.

The other point of concern is that the rapid investment in CCGTs has proceeded without any planning for capacity needs, based solely on the aim to promote competition and short-term profit considerations of private firms. Any potential generator has been able to receive a licence. Clearly, it would have made both environmental and economic sense to explore the possibility of some demand-side management, rather than rapidly building up such a large new capacity. This could have been done within a framework of liberalisation, as examples of competitive bidding systems from the US show. This opportunity has now essentially been lost for the foreseeable future, as it is unlikely that much new capacity will be required in the near future as demand is growing only very slowly, around 0.5 per cent per annum.

Nuclear's demise and support for renewables

While gas has been a winner, nuclear power has fallen foul of the profitability requirements of a privatised system, although even before privatisation the expansion of the nuclear programme had encountered problems. While initially, all nuclear plants were taken out of the privatisation programme and supported through a fossil fuel levy (around 10 per cent of end-user prices), linked to an obligation on the RECs to purchase nuclear power (through the Non-Fossil Fuel Obligation – NFFO), it was subsequently decided that the more profitable plants were to be privatised in 1996. However, in the event 12.3 per cent of shares had to remain in public ownership, as not enough investors showed interest, the first time that a UK privatisation offer was not oversubscribed. The new company, British Energy, has announced that it has abandoned plans to build any further nuclear plants, and is expected to invest in CCGTs. The UK experience has shown that nuclear power, in particular with its high capital as well as decommissioning costs, is currently not a preferred option in a liberalised energy market.

In CO_2 terms, the demise of nuclear power is essentially a negative development. Most of the currently operating plants will reach the end of their design life between 2000 and 2010. Only the Sizewell B PWR, which was commissioned in 1994, will operate for a substantial period past that time. Hence, once gas has reached its maximum contribution to generation (to which there are limits related to gas prices), further CO_2 reductions are difficult to achieve. The overall environmental credentials of nuclear power are less clear, in view of accidental risks and unresolved issues regarding waste disposal, so that many environmentalists have welcomed the demise of nuclear power.

More clearly positive has been the situation regarding renewable energies. Since privatisation, the situation for renewable energies has improved drastically, mainly as a result of their inclusion in the NFFO and the fossil fuel levy, which were initially conceived to protect nuclear power. The government established a 600 MW quota for renewable energies, to be filled in stages by 2000, as a means of achieving an overall government target of 1500 MW of renewable energy by that date. Three separate orders for renewable projects under the NFFO have been made, contracting 388 schemes totalling 1343.27 MW. By mid-1996, 351.9 MW were operational.

It seems possible that the target of 1500 MW of renewable energy by 2000 will be achieved. However, this figure is still a small proportion (4.6 per cent) of the current UK generating capacity. Mitchell (1995) argues there has been too much emphasis on costs and not enough on environmental impacts, diversity and longer-term options. Nevertheless, despite some shortcomings, the NFFO has much improved the chances for renewable energy in the UK and renewables are effectively set to replace nuclear power as the primary non-fossil fuel source in power generation within the next twenty or so years. Apart from government financial support, which could also have been provided under a nationalised, non-liberalised system, the more competitive electricity sector regime has been important in the development of renewables. The RECs, for example, see investment in renewables as one option to achieve more independence from the generators and to expand the unregulated side of their business, although only under subsidy conditions. Supply competition may also benefit renewables, as there are indications that customers might be interested in purchasing 'green electricity' at a premium from renewable energy companies, which is already happening in the Netherlands. In the UK, one renewable energy company has received a licence but interest might grow when the market opens to domestic customers in 1998.[9]

Overall, therefore, on the generation side, electricity privatisation and liberalisation has had some beneficial environmental effects, although they are to some extent a result of the specific regulatory framework that has been established, rather than liberalisation per se. Furthermore, the generation side is only part of the story. It is generally accepted that a main focus of a more sustainable energy policy has to be energy efficiency.

End-use efficiency – some small incentives

Electricity privatisation provided the government with a number of opportunities to move towards a more efficient energy system. Examples from the US show that private utilities can be coerced into forceful DSM provided there are regulatory provisions. The government's sustainable development strategy clearly states that energy markets should operate within frameworks which do not undermine efforts to improve energy efficiency (HM Government, 1994). However, the evidence so far is that liberalisation has done little to promote energy efficiency and may result in doing the opposite. A particular problem has been the lack of consideration given to energy efficiency by OFFER and OFGAS, as a result of inadequate regulatory provisions.

The gas privatisation bill contained no requirement for the promotion of energy efficiency, while the Electricity Act stated that the Secretary of State and DGES shall have the duty to 'promote efficiency and economy on the part of persons authorised by licences to supply or transmit electricity and the efficient use of electricity supplied to consumers'.

Additionally, the DGES may 'determine such standards of performance in connection with the promotion of efficient use of electricity by consumers as, in his opinion, ought to be achieved by such suppliers' (Electricity Act, 1989).

Such standards were set by OFFER in 1994 but they are relatively unambitious. The required savings amount to just over 2 per cent of current supply. Meanwhile, the government made energy efficiency the supposed cornerstone of its climate policy and for this purpose set up the Energy Saving Trust (EST) in 1992. The Trust was set up as a joint partnership between the government, British Gas, the RECs, Scottish Power and Scottish Hydro Electric. The aim was to develop, propose and manage programmes to promote energy efficiency, so as to stimulate markets. To achieve its targets, the EST was initially expected to stimulate and coordinate up to £2 billion worth of financial incentives, loan options and customer advice by the year 2000. It started with a budget of under £5 million in 1993/94 and was supposed to reach an investment profile of £400 million by the year 2000. Most of this funding was to come from surcharges on gas tariff and electricity franchise customers. By 1997/98 it was expected that finance from those two areas would reach £150 million each (Energy Saving Trust, 1994).

Initial support came from the gas regulator who imposed a small consumer levy, the so-called e-factor, to finance EST pilot projects. Then OFFER made some analogous concessions to energy efficiency in the shape of a revenue allowance, linked to the energy efficiency standards of performance, which was implemented on 1 April 1994, to run until 1998. A figure of £1 per franchise customer per annum for four years has been set (amounting to around £100 million overall over 4 years), with projects to be approved by the EST. Meanwhile, OFGAS in 1994 decided to cancel the e-factor, leaving the EST with an immense funding shortfall.

Furthermore, as of 1998, supply will be totally liberalised and the assumption is that the market will take care of energy efficiency then. Diesendorf (1996) has argued to the contrary in the case of Australia, where he feels that a competitive retail market would severely discourage efficient energy use. In the UK, it is true that the RECs are to some extent using energy efficiency as a selling point for industrial customers, for example by running technology demonstration centres, although their purpose is mainly to encourage customers to switch from other fuels to electricity. Under the current regulatory system, profits are linked to sales. In principle, energy efficiency could be used to attract customers when the market opens up totally. However, the RECs are dubious whether this would be a cost-effective way for attracting customers and feel that it would probably not be worthwhile in the small customer market.

One of the greatest obstacles to energy efficiency in the liberalised UK energy market is to do with energy prices. The 1994 distribution price controls involved initial price reductions of between 11 and 17 per cent, followed by further reductions of 2 per cent a year in real terms. Further cuts were announced in 1995 and customers were given a rebate of £50 as a result of the sale of the National Grid Company. With the sale of British Energy the fossil fuel levy has been abolished, resulting in a further 10 per cent cut in prices. As already mentioned, low energy prices are universally recognised as a major disincentive to investments in energy efficiency. As, because of low energy prices, consumers have no great incentive for making energy efficiency improvements, supply companies also can see no great value in employing energy efficiency as a marketing tool. Liberal-isation as such is thus unlikely to improve the prospects for energy efficiency.

LOCAL ENERGY COMPANIES IN GERMANY – ENVIRONMENTAL PERFORMANCE UNDER THREAT?

As the previous section has shown, liberalisation in the UK has had rather mixed results in environmental terms. Meanwhile, in some non-liberalised systems (e.g. Denmark, Germany and Sweden), there are examples of energy companies with ambitious environmental objectives. This section will discuss the case of local energy companies in Germany and the potential implications of liberalisation.

The structure of the Germany energy sector is rather complex (see Collier, 1994a). From an environmental point of view, the activities of a number of local energy companies are most interesting. While all local authorities own distribution networks for gas and electricity, many have leased supply rights to the large regional energy companies. However, a considerable number, especially (although not exclusively) in the larger towns and cities, have set up municipal energy companies, the so-called *Stadtwerke* (often also responsible for water and local public transport). In most cases they generate at least part of a town's electricity requirements. It is here that a number of environmentally inspired schemes can be found.

Originally, as Lutter (1990) establishes, the driving force for many local energy programmes was a political rejection of nuclear power (which still continues despite climate change concerns), high oil prices (as the heating sector in Germany was dominated by oil), and environmental concerns (in particular acid rain). Climate change has become an additional priority during the last five years and has resulted in some new initiatives. While the exact focus of local energy programmes varies, some common characteristics can nevertheless be identified:

- plans and programmes are drawn up jointly between local authorities, local councillors and the energy company;
- preference for CHP power plants (coal and gas);
- integrated heat planning with priority areas for district heating;
- energy efficiency programmes with subsidies.

The energy programmes in three cities are briefly discussed here.

Three best-practice examples

Saarbrücken probably has one of the best known municipal energy programmes in Germany and has won a number of prizes, including one at the Rio summit. The *Zukunftskonzept Energie* was originally conceived in 1980, with a major focus on the heating sector (expansion of district heating and gas). Interesting in climate change terms is the commitment to utilise coal as a local resource, but to reduce emissions through efficient CHP generation. Additionally, waste furnace gas and methane gas from local coal deposits are used for power generation. A subsidy scheme for photovoltaics also exists, aiming at the installation of 1000 kW (0.5 per cent of overall demand) by 2000. Furthermore, there are various energy efficiency information activities, as well as grant and loan schemes. These measures have resulted in substantial emissions reduction, of 76 per cent of SO_2 emissions, 34 per cent of NO_x and 15 per cent of CO_2 between 1980 and 1990 (Stadtwerke Saarbrücken, 1991). However, the achievement of a further 25 per cent reduction of CO_2 emissions by 2005, set as a goal by the city council, is proving a challenge. Particular problems are low energy prices, the potential threat of competition from France under the Internal Energy Market, municipal budget constraints and the lack of an appropriate regulatory framework to enable more extensive DSM.

Hannover has also achieved much in the past and is particularly interesting in that it recently commissioned five consultants to carry out a least-cost planning study (with partial funding under the EU SAVE programme) as a means to determine the possibility for further energy/CO_2 savings. This calculated a savings potential of 6 per cent of electricity generation compared to the reference scenario and a CO_2 reduction of 4 per cent (Stadtwerke Hannover, 1995). A main problem for the realisation of these scenarios is that while there are macroeconomic benefits (both with and without environmental cost internalisation), the *Stadtwerke* would actually have to sustain an annual loss of 242 million DM. This represents 3 per cent of annual turnover and would have to be financed by electricity price increases. However, the *Stadtwerke* feel that in an increasingly competitive market they could not afford to impose such increases. In fact, they might not be approved by the regional price control authorities. Meanwhile, the *Stadtwerke* have proposed the establishment of a 'climate change fund' providing 10 million DM per annum to finance DSM programmes.

Leipzig exemplifies the rather different situation in the new Länder. There, the previously state-owned energy companies were to be carved up between the large Western German regional energy companies, according to the *Stromvertrag* (Electricity Contract) of 1991. However, this was contested before the constitutional court by 164 authorities and a compromise agreement awarded them the right to form their own *Stadtwerke* (Collier, 1994a). Some of these, such as Leipzig, have now drawn up environmentally focused energy programmes, with the emphasis on the reduction of air pollution. In the new Länder, the heating sector was already dominated by district heating although this was generally supplied by heat only plans, transported in uninsulated, overground pipelines. Priorities are now the renovation of these networks, the construction of efficient CHP plants and a range of energy efficiency measures, for which there are some federal subsidies. Leipzig has built a 172 MWe CCGT CHP plant with plans for a second one. This, together with other measures and a substantial reduction in energy use through industrial restructuring, is expected to result in CO_2 emission reductions of over two-thirds by 2000.

Liberalisation as a threat

It has to be recognised that these three cities are amongst the leaders in Germany as far as environmentally compatible energy systems are concerned. Not all German *Stadtwerke* are that advanced. Nevertheless, these three examples illustrate what can be achieved, if the political will exists. Within this context, it is important to stress that, although local energy concepts are affording priority to environmental concerns, this has not been entirely at the expense of economic considerations. One reason why many towns have increased their own generation capacities is to gain more independence from the large regional utilities and make profits from electricity generation. However, many decisions have been influenced politically and certain non-profitable schemes have been promoted for environmental reasons (e.g. the 1000 kW PV programme in Saarbrücken). Over recent years, local authority budget constraints have set limits to such activities, as the *Stadtwerke* have come under increased pressure to produce high returns. Financial constraints could also threaten energy efficiency programmes as they mean fewer sales.

The main point to make here is that these local energy concepts depend very crucially on the monopoly position of the *Stadtwerke*, as well as the lack of competition between district heating and gas. It is quite clear that district heating could not have expanded to the same degree had there been straight competition between gas and heat. The designation of priority areas for district heating has been of crucial importance. Also, district heating allows the *Stadtwerke* to build relatively expensive generating plant. This supposedly inefficient subsidisation has been criticised by the German monopoly authorities.

The *Stadtwerke* are seriously concerned about the prospects of liberalisation. While in principle they could benefit, for example, by being able to sell services to neighbouring areas who are currently supplied by the regional energy companies, the most likely scenario under a more liberalised system would be the large, powerful energy companies attempting to undercut the *Stadtwerke*. Few of them feel they would be able to compete, especially in the larger customer bracket. Thus their profits would be eroded and they would have even less money for measures like energy efficiency programmes which are only marginally profitable.

Another important aspect of the *Stadtwerke*'s operation is the cross-subsidisation of public transport services, either directly or indirectly through profits from energy operations which flow into the municipality's budget. Obviously, environmentally, good public transport is extremely important. In a liberalised system, reduced profits of the *Stadtwerke* would result in less money being available for public transport and other measures taken by the local authority.

Overall, liberalisation can thus be considered a substantial threat to the activities of municipal energy companies. Some of the negative aspects may be compensated for by some of the potentially positive aspects of liberalisation, but this cannot be taken for granted.

ASSESSMENT AND CONCLUSIONS

In environmental terms no generic assessment can be made about the respective benefits and disadvantages of liberalisation versus the operation of public or private monopolies, as much depends on specific circumstances. The two opposing cases of the UK and Germany have highlighted some of the potential benefits and disadvantages. Obviously, two case studies do not allow a full assessment but nevertheless provide some broad indications.

Neither system has environmental benefits per se. In the UK, emission reductions are currently resulting on account of:

- the economic attractiveness of CCGTs;
- regulatory measures to support renewables and energy efficiency.

In Germany, on the other hand, emission reductions are a result of:

- political priorities pushing for emission reductions, enabled by political influence on the operation of energy companies;
- the economic attractiveness of CHP plants, partially as a result of integrated heat planning;
- the service orientation of municipal energy companies.

In the liberalised and privatised UK system, economic viability and profit maximisation are an absolute priority. In Germany, although the *Stadtwerke* are expected to make profits for their municipal shareholders, as a municipally owned company they are also expected to meet social and environmental objectives. There is both direct political control and influence through the presence of councillors on company boards. Economists generally criticise this type of political influence as interference with the market. However, there is no evidence that this type of intervention is necessarily any worse than other forms of regulatory control.

In the UK, regulatory changes could result in a better environmental performance by the new system, especially in the long term. It is quite clear that the market will not supply energy efficiency by itself. Incentives must be created for energy companies to pursue DSM programmes. Compulsory competitive bidding for DSM along the model of some US states would be one possibility, thus effectively creating a market for energy services (see also Diesendorf, 1996). A 'free for all' on the generation side may deliver a balance between generation plant and DSM investments in the long run, but has so far mainly resulted in generation overcapacity. The competition from natural gas makes CHP (both large- and small-scale) a difficult proposition. This is unlikely to change unless local authorities receive some powers to designate DH priority areas, as in the case of Germany. For renewable energies, a liberalised system is essentially a positive development, although in the short to medium term, public subsidies will continue to be needed.

In both systems, the preoccupation with low energy prices conflicts with emission reduction objectives, in that it even further reduces incentives for energy efficiency. While industrial competitiveness is obviously an important issue, low energy prices are not necessarily a crucial determinant, as the frequently quoted example of Japan exemplifies. In any case, there is a need to shift the discussion from concerns about low energy prices per unit to a consideration of total energy bills (or costs per energy service). In many industrial sectors, these can be significantly reduced through energy efficiency investments but new incentives have to be created to encourage them. While little progress is made with energy/carbon taxes, regulatory authorities have other possibilities, such as tariff surcharges (akin to the now abandoned e-factor in the UK gas price control) or encouraging 'profit sharing' for energy efficiency investments.

A main conclusion to be drawn out of the discussion in this chapter is that a liberalised system can produce some environmental benefits but needs a tight regulatory framework to maximise them. At the same time, liberalisation can present a threat to local energy companies which operate according to a balance of economic, social and environmental aims. In such a system, a wider definition of efficiency must be employed before any steps towards liberalisation are taken, especially when the prospects for cost internalisation measures (such as carbon taxes) look remote.

Taking the EU as a whole, there is little doubt that a number of changes to the energy sector are desirable on environmental grounds. There is a need for a greater consideration of demand-side investments almost everywhere (except possibly in Denmark and Sweden). Considering the varying situations at present, there is no one organisational or regulatory model that can be recommended, but it is imperative that energy sector reform does not proceed on the basis of narrow economic considerations alone.

NOTES

1 This chapter draws on interviews with energy company representatives and local authority officials in March–July 1995 (for the UK) and December 1995 (for Germany), as part of the project 'Climate Change Policies in the European Union', carried out at the European University Institute between October 1994 and March 1996. Co-financing from DG XI of the European Commission is gratefully acknowledged.

2 For more detail on electricity sector structure see OECD (1994).
3 As reported in *Stromthemen*, 7/96, p. 3.
4 It is beyond the scope of this chapter to explore the microeconomic and social issues associated with liberalisation. In the UK, electricity sector privatisation and liberalisation has brought with it immense changes in the market shares and profitability of individual companies, as well as large job losses.
5 For more details on electricity privatisation, see Eikeland (1995) and Surrey (1996).
6 It is beyond the scope of this paper to go into any detail on the new structure, for more detail see e.g. Department of Trade and Industry (1995).
7 At the time of writing (November 1996), the most recent emission figures available were for 1994.
8 However, National Power and PowerGen have at times used other, more polluting plant in preference to the FGD equipped plant as it increases generating costs.
9 As reported in *ENDS Report* 254, March 1996, p. 27.

REFERENCES

Arentsen, M.J. and Künneke, R.W. (1996) 'Economic organisation and liberalisation of the electricity industry', *Energy Policy* 24 (6), pp. 541–552.
Bishop, M., Kay, J. and Mayer, C. (1995) *The Regulatory Challenge*, Oxford: Oxford University Press.
Boehmer-Christiansen, S. and Skea, J. (1991) *Acid Politics*, London: Belhaven Press.
Bonbright, J., Danielsen, A. and Kamerschen, D. (1988) *Principles of Public Utility Rates*, Arlington: Public Utilities Reports.
Collier, U. (1994a) 'Local energy concepts in Germany: an environmental alternative to liberalisation?', *Energy and Environment* 5 (4), pp. 305–326.
—— (1994b) *Energy and Environment in the European Union: the Challenge of Integration* Aldershot: Avebury.
Collier, U. and Löfstedt, R. (1997) 'Think globally, act locally? Local climate change strategies in Sweden and the UK', *Global Environmental Change* 7 (1), pp. 25–40.
—— (eds) (1997) *Cases in Climate Change Policy: Political Reality in the European Union*, London: Earthscan.
Department of the Environment (1996) *Digest of Environmental Statistics No. 18 1996*, London: HMSO.
Department of Trade and Industry (1995) *Energy Projections for the UK*, London: HMSO.
—— (1996) *Digest of UK Energy Statistics*, London: HMSO.
Diesendorf, M. (1996) 'How can a "competitive" market for electricity be made compatible with the reduction of greenhouse gas emissions?', *Ecological Economics* 17, pp. 33–48.
Eikeland, P.O. (1995) 'Norway and the UK: a comparative institutional analysis of competitive reforms in the electricity supply industries', EED Report 1995/3, Lysaker: Fridtjof Nansen Institute.
Electricity Act (1989) London: HMSO.
Energy Saving Trust (1994) *Strategic Plan 1993–2000*, London: EST.
European Commission (1995) 'White Paper: an energy policy for the European Union', *COM* (95) 682 final.
Flavin, C. and Lenssen, N. (1994) 'Reshaping the electric power industry', *Energy Policy* 22 (12), pp. 1029–1044.

Hirst, E., Cavanagh, R. and Miller, P. (1996) 'The future of DSM in a restructured US electricity industry', *Energy Policy* 24 (4), pp. 303–315.
Jaccard, M. (1995) 'Oscillating currents: the changing rationale for government intervention in the electricity industry', *Energy Policy* 23 (7), pp. 579–592.
Kozloff, K. and Dower, R. (1993) *A New Power Base – Renewable Energy Policies for the Nineties and Beyond*, Washington: World Resources Institute.
Löfstedt, R. (1997) 'Sweden: the dilemma of a proposed nuclear phase out', in Collier, U. and Löfstedt, R. (eds) op. cit.
Lutter, H. (1990) 'Zehn Jahre Erfahrungen mit örtlichen und regionalen Energiekonzepten in der Bundesrepublik Deutschland', *Informationen zur Raumentwicklung* 6/7, pp. 305–314.
Majone, G. (1994) 'Paradoxes of privatisation and deregulation', *Journal of European Public Policy* 1 (1), pp. 53–69.
Mitchell, C. (1995) *Renewable Energy in the UK – Financing Options for the Future*, London: CPRE.
Nadel, S. and Geller, H. (1996) 'Utility DSM. What have we learned? Where are we going?', *Energy Policy* 24 (4), pp. 289–302.
OECD (1994) *Electricity Supply Industry: Structure, Ownership and Regulation in OECD Countries*, Paris: OECD.
Rüdig, W. (1986) 'Energy conservation and electricity utilities', *Energy Policy* 14 (2), pp. 104–116.
Sioshani, F. (1996) 'DSM in transition: from mandates to markets', *Energy Policy* 24 (4), pp. 283–284.
Stadtwerke Hannover (1995) Integrierte Resourcenplanung: Die LCP Fallstudie der Stadtwerke Hannover AG Ergebnisband, Hannover: Stadtwerke.
Stadtwerke Saarbrücken (1991) Das Saarbrücker Zukunftskonzept Energie, Saarbrücken: Stadtwerke.
Surrey, J. (1996) *The British Electricity Experiment*, London: Earthscan.

7

LIBERALISATION OR DEREGULATION? THE EU'S TRANSPORT POLICY AND THE ENVIRONMENT

Michael Teutsch

INTRODUCTION

Governments have a long tradition of intervening in transport markets. This is because most transport systems need some kind of rule-setting in order to work properly. For example, a rule establishing on which side of the road traffic moving in the same direction should use, makes life easier, and most probably longer as well. In addition to such coordination problems, transport policy is concerned with the distribution of public goods, which might also be a prerequisite for private wealth. Transport policy has an impact on regional development, the availability of services for certain social strata, or the competitiveness of specific industries. Furthermore, transport is a source of accidents, noise, and exhaust fumes. Therefore, the avoidance of negative external effects is a third reason for public activity in this policy field.

Although agreement might be reached on the existence of problems related to coordination, distribution, and avoiding negative external effects, opinions are certain to diverge widely on the best means to be applied in order to deal with them. For a long time attention has been drawn to the problem of market failure in the transport sector. Consequently, economists as well as politicians have called for corrective state intervention and legitimised existing regulative systems. However, the extent to which governments have intervened in transport markets has varied considerably. Whereas some countries have limited their intervention to the absolutely necessary and have left the rest to be coordinated by market forces, others have substituted market mechanisms, e.g. by imposing restrictions to market access and administrative price-regimes. In addition, many countries have not limited transport policy to the regulation of transport systems as such; instead, they have frequently used it to help accomplish goals of regional, social, industrial, or defence policy. State-owned railways have played an important role herein, as they have given the government an important policy instrument that could be used to interfere directly in the operation of transport markets.

However, in recent years there has been a growing awareness of government (or intervention) failure in transport markets. In the wake of a deregulatory drive in many industrial and service branches, the extent to which transport markets actually need regulation in order to work properly has been questioned increasingly (cf. Button, 1992; Barde and Button, 1990). The EU framework has played an important role in liberalising international transport and, indirectly, in changing some of the pre-existing domestic regulative systems in the Member States.[1] The Treaty of 1957 asked the Six to develop a common policy in the field of transport, especially inland transport. It called for the liberalisation of international transport services, as well as for 'additional measures' that should be taken to make transport markets work properly. However, the development of a common transport policy was deadlocked for decades, due to conflicts between high-regulating Member States and those advocating a quick liberalisation of transport on the other hand. This situation was only recently resolved in one of the most important fields of activity, namely road haulage. The liberalisation of international road transport services in the Community, as well as the opening of national markets for non-resident hauliers were finally agreed in 1988 and 1993 respectively.

Apart from actors resisting this development because they had profited from protected markets for decades and now feared hard competition with foreign service industries (cf. BDF, 1994), there are also concerns about the impact of deregulation on the level of environmental degradation caused by the transport system (e.g. Hey, 1994). These critics are worried about high increases in traffic induced by liberalising trade, as well as by liberalising transport markets. A main concern relates to the disproportionately large increase of road and air transport. The latter have become the dominant modes of transport, even more on international than on domestic markets and are generally regarded as the most environmentally damaging means of transport.

One of the central features of this chapter is to contrast deregulation with liberalisation. Whereas the two are very often used as synonyms, a clear distinction must be drawn between them. While deregulation simply abolishes restrictions to the application of market rules, liberalisation (or regulatory reform, cf. Button, 1991: 1), as well as creating markets, ensures their proper working by introducing additional rules. It will be argued that abolishing some of the existing regulative instruments in the EU Member States should not necessarily be considered negative with respect to the environmental impact of transport systems, as long as the appropriate accompanying measures are taken, in particular the full internalisation of costs. The chapter will then discuss EU policies regarding transport and the environment, and present some political and institutional arguments to explain the EU Commission's emphasis on liberalisation and the use of fiscal instruments in dealing with the topic of transport and the environment.

TRANSPORT AND THE ENVIRONMENT

During the last decades transport has become one of the most important sources of air pollution and a major cause of noise nuisance. It accounts for about 60 per cent of carbon monoxide emissions and for 25 per cent of energy-related carbon dioxide emissions. Road transport is the source of 80 per cent of all transport-related carbon dioxide emissions, more than 50 per cent of nitrogen oxides (NOx) emissions and a major part of volatile organic compounds (Commission, 1995b: 1). In addition to the overall increase in transport activities, the modal split has undergone alarming developments. There has been a rapid growth of road and air transportation, while the generally more environmentally friendly transport modes – railway and inland waterway shipping – lost a great part of their former importance. In 1992, road haulage already held a 70 per cent share of European freight transport markets (measured in tonne/kilometres), 16.3 per cent of all transport activities were realised by rail transport while inland waterway shipping accounted for 4.4 per cent. Forecasts predict a doubling of transport demand between 1988 and 2010 with particularly high growth rates in international transport. Assuming that the current economic, technological, and political framework does not change considerably, the trend towards a further strengthening of the position of road haulage and aviation is likely to continue (Commission, 1995b: 1, cf. DIW, 1994).

When discussing options for making transport more compatible with the concept of sustainable development, a number of different strategies can be taken into consideration (cf. Enquete-Kommission, 1994: 123ff.).

The first would aim at an overall reduction in transport volumes. As the development of transport demand is usually seen as deriving directly from economic growth and the spatial distribution of economic activities, it clearly is very difficult to achieve progress in this domain. Legitimate interests in economic and regional development are sure to provoke strong opposition against measures which appear to aim at the reduction of transport volumes at the expense of the production of wealth. In addition, efficient ways of land use planning, which could take into account these problems and reconcile economic as well as environmental needs, presuppose a high intensity of coordination and will only be effective in the long run.

A second option is to ensure that existing transport markets function in such a way that their environmental impact is minimised. This would include the use of the best available technologies, continuing technological

116

development to reduce pollution and energy consumption of the engines, and making more efficient use of existing capacities, e.g. increasing load factors.

Thirdly, transport volumes can be shifted to less environmentally damaging transport modes. However, it is difficult to make an exact assessment of the specific energy consumption of the different modes of transport, as a multitude of factors influence this variable, such as load factors, transport distances, and the number of transfers to other means of transport. According to a European Commission Green Paper on Transport and the Environment (Commission 1992a: Figure 3) rail and inland waterway transport are the most energy efficient means of transport (0.6 MJ primary energy per tonne km). However, in the case of the railways this is only true for the transport of bulk goods. Wagon load shows a higher figure (1.0 MJ/t–km), which is comparable with a 5 axle articulated 38 ton-truck, when an average load factor of 70 per cent is assumed (0.99 MJ/t–km). With an (unrealistic) load factor of 100 per cent such a heavy truck can even arrive at 0.69 MJ/t–km, whereas its specific energy consumption rises to 1.38 MJ/t–km when a 50 per cent capacity utilisation is presumed.[2] Generally speaking, smaller and rigid lorries show higher figures of specific energy consumption.

The interpretation of such comparative figures is further complicated by the difficulty in taking into account the different character of service that the different means of transport typically provide: for instance the difference between short-distance transport of relatively light, but high-value goods on the one hand and the transport of bulk goods covering large distances on the other hand. The environmental problems connected to rising market shares of road haulage also derive from developments of transport markets which are unlikely to be responsive to any kind of political intervention. Transport markets have seen a trend towards a greater number of relatively light goods which are typically transported by road hauliers. Consequently, the market share of railways and inland navigation, which were best suited to transporting heavy goods such as coal and steel, diminishes as long as no action is taken to reverse this trend. However, technical and logistical difficulties may increase the difficulties in shifting goods from one means of transport to another.

TRANSPORT DEREGULATION

What does deregulation in the field of transport mean and what are its potential environmental implications? In order to answer this question one first needs to assess the practical experience with regulation at the national level. The following questions have to be answered. What types of fiscal and regulative instruments have been used in the transport sector? What have been the regulatory goals? And finally, what has been the outcome in terms of the functioning of markets, the achievement of the envisaged goals, and the environmental impact of transport activities?

Most countries in continental Europe (the UK's liberal approach towards transport policy has been the exception in Europe for a long time (cf. Button, 1991)) have applied some control over their national markets until recently. For the most part, regulation was introduced during the economic crisis at the beginning of the 1930s and has aimed at ensuring market stability. Markets were meant to be protected against phenomena such as cut-throat and unfair competition. Last but not least, in countries such as Italy, France and Germany the protection of the state-owned national railways was another important goal lying behind regulation of inland-waterway shipping and especially the fast growing road-haulage sector. The most distinctive instruments used have been the quantitative limitation of concessions for road hauliers and the application of fixed rates for transport services. Generally speaking, Germany has been the country with the strictest regulative regime in terms of legislation and implementation. Thus, to become a road haulier it was not sufficient to meet a number of personal criteria. Permission also depended on numerical restrictions which were independent of a candidate's personal standing. As far as pricing mechanisms for transport services are concerned, rates were under the control of transport authorities and orientated at rail transportation tariffs for a long time. They could not be negotiated freely between individual shippers and operators, but were set by special committees, transport ministries, etc.

What was the outcome of this kind of regulation? First of all, the regulation of road haulage has failed completely to meet the envisaged target of protecting the railways and to impede the growth of road transport. Secondly, price and capacity regulation may have provided transport markets with some stability. But they have also protected from further competition those who already were in the market and have led to relatively high price levels when compared to non-regulated markets. According to the views dominating transport economics today, this stability has meant preserving outdated market structures, and the slowing down of technical and organisational innovation processes (Hamm, 1989; Deregulierungs-kommission, 1991). Firms of doubtful efficiency and competitiveness were able to stay in the market longer than they would have done under conditions of free competition. Thus regulation has impeded the flexible and most efficient utilisation of transport resources. It has led to avoidable transport operations and, consequently, also to avoidable noise, exhaust gases, etc., which is an unnecessary waste of resources (both economic and environmental) par excellence.

To give an example: comparison with the development of the US trucking industry after deregulation in 1980 (cf. Chow, 1991) shows that the highly regulated systems of Germany, as well as Italy, have tended to preserve existing market structures and the predominance of small and medium enterprises. Even if the preservation of small and medium-size enterprises is sometimes regarded as a policy goal by itself, it should be borne in mind that smaller transport firms may have greater difficulties in improving their efficiency by means of introducing modern disposition and logistics systems. According to a recent empirical study, they usually achieve lower load factors than larger enterprises (Baum and Sarikaya, 1995: 117).

The same is true for own account transport when compared with hire and reward haulage (Baum and Sarikaya, 1995: 117). Restrictions to hire and reward hauliers have usually been accompanied by a rise of own account transport. That is, a growing number of industrial enterprises, for example, built up their own transport capacities, as own account transport has usually not been subject to quantitative restrictions in any country. But as a consequence of organisational problems, as well as of rules such as those still found in Germany, which strictly forbid a half-loaded truck belonging to one firm to carry goods for another firm, even when they belong to the same group, own account transport does not use its capacities in the most efficient way and thus contributes to a waste of resources. In conclusion, regulation which gives incentives to shippers to build up their own (less efficient) transport capacities and constrains them to work in a less than optimal manner is an example of an economically and environmentally undesirable form of intervention into transport markets.

It must be admitted that the efficiency argument, which is used here, has been contested frequently, as some authors fear excess capacity and increasing numbers of empty or half-empty trucks on the roads as a consequence of the abolition of quantitative restrictions to market access (cf. BDF, 1995b). In addition, deregulation that results in a further fall of prices for road transport can ultimately damage the competitive position of more environmentally friendly modes of transport (Hey, 1994: 91; Baum *et al.*, 1990: 31ff.). The abolition of mandatory tariffs in Germany at the beginning of 1994 brought about an abrupt fall of prices for road transport on the domestic market, where they had been estimated to be about 20 per cent higher than for comparable services in liberal regimes (e.g. international road transport, Hamm, 1989: 26; Deregulierungskommission, 1991: 46f.; cf. BDF, 1994, 1995a). It is difficult to determine whether such phenomena constitute short-term distortions resulting from adaptation processes, or whether they are of a systematic nature. In any case, this argument does not justify the suspension of market mechanisms in the price setting strategies of those who actually demand and supply transportation services. The discussion should rather focus on another type of instrument, namely on correct pricing for the use of natural and infrastructure resources and on taxation.

GETTING THE PRICES RIGHT: THE PROBLEM OF EXTERNAL COSTS

Correct pricing mechanisms are not only relevant for the efficient calculation of costs inside a certain sector, but are also an indispensable prerequisite for fair competition between different modes of transport. Consumers choose a particular means of transport on the basis of its price and its quality. As has been said before, there should be sufficient economic incentives at the individual level to use transport capacities in the most efficient way. Pricing should aim at balancing the economic rationality of an individual haulier or transport user with the rational use of resources at the societal level. Thus, all costs arising from operational practice and all kinds of scarcities should be reflected in prices. However, there are effects produced by transport operations which are currently not reflected in markets and, more specifically, the prices that are to be paid. These external effects, such as the environmental costs of transport, welfare losses due to congestion and accidents, are unlikely to be taken into consideration by consumers before opting for a particular transport option (cf. Commission, 1995b; ECMT, 1994).

External costs arise, as there is no natural actor directly representing environmental or similar interests in the economic system. Hence, it inevitably is the responsibility of the state to deal with these costs. Policy must impose upon the market the reflection of environmental and other external costs. The usual mechanism to do this is by introducing taxes or certain charges for the use of infrastructures and natural resources, which should represent incentives for transport users to check their real transport needs and to demand less environmentally damaging services (cf. Crawford and Smith, 1995; Baum and Sarikaya, 1995).

Numerous economists have tried to provide estimates of the external costs of transport and to present them in monetary terms (cf. Commission, 1995b). However, there still is broad and sometimes heated discussion on methodological problems in defining and quantifying external costs. The questions that are raised concern, for example, the criteria applied for the definition of certain costs as external and the fiscal quantification of environmental damages, damages to persons caused by accidents and welfare losses due to congestion. Finally, the question arises whether external benefits are produced contemporaneously to external costs and if the former could be taken to offset the latter (cf. ECMT, 1994).

This rather high level of scientific uncertainty leaves the political process open to the influence of specific economic, sectoral, and regional interests. The general decision to introduce environmental taxes, the choice between different fiscal instruments, and the exact fixing of certain levels of taxation is of course subject to political decision-making and cannot be deduced simply from economic models. Thus, the institutional structure of political systems and the interest conflicts that take place within these polities have to be taken into account when it comes to assessing whether environmental taxes for transport are likely to be introduced on the EU or on the Member States level in the near future. The next section thus tries to describe how the problem has been approached in European legislation up until now.

EU POLICIES REGARDING TRANSPORT AND THE ENVIRONMENT

For decades, the main focus of EU transport policy had been the creation of a common transport market and, as a consequence, the harmonisation of legislation which might distort competition, such as taxes or technical and social regulation. Recently, the Commission has tried to extend the scope of the Community's transport policy beyond the goal of market-creation. Among other things, its new approach puts a greater emphasis on ensuring that European transport markets work without damaging the interests of third parties such as the general public. Thus, the Commission has made considerable efforts to develop policies which aim at reconciling transport and environmental needs. It published a Green Paper on transport and the environment in 1992 (Commission, 1992a). The White Paper on the future development of the common transport policy of the same year put great emphasis on the concept of 'sustainable mobility' (Commission, 1992b). In 1995 followed a Green Paper on the external costs of road transport (Commission, 1995b). Finally a White Paper on the development of the Community's railway policy was published in 1996 (Commission, 1996a).

The 1992 White Paper presents two basic strategies for reducing the environmental impact of transport, which have also been mentioned in section two. First, reducing operational pollution and second, a demand-side centred approach, that is making more efficient use of existing capacities and shifting transport volumes to more environmentally friendly means of transport. Although agreement on reducing the operational pollution of transport is likely to be reached more easily,[3] it is not regarded as sufficient to reduce the overall environmental damage caused by transport. This is because all improvements in the technical field are likely to be more than compensated for by the expected rise of transport volumes in passenger, as well as in freight transport (Commission, 1992b: no. 173). Therefore, alternative strategies and their political implications have to be discussed as well.

With regard to shifting goods between different modes of transport, there are, generally speaking, three theoretical options for doing this:

• Limiting the consumers' freedom of choice for transport services through administrative means. This strategy is, however, likely to provoke strong resistance by transport users as well as by the discriminated operators. It can thus only be achieved at high political cost.

- Transport users' choices can be influenced by raising the prices of environmentally damaging modes of transport. This alternative has the advantage that apart from providing incentives for the use of alternative modes, it could also help to reduce the overall demand for transport. It can be assumed that opposition to this option will also be strong, because it increases transport costs. But compared to the first option resistance is likely to be less pronounced, as the consumer's freedom of choice is still respected in principle.
- Making environmentally friendly means of transport more attractive by improving their quality of service. This is likely to meet the least resistance because there is neither a direct limitation to the freedom of choice of the consumer, nor are transport prices touched directly. The redistributive goals of policy-makers are least obvious and transport consumers can only win from improvements in service quality.

When discussing the development of transport policies at the EU level, the kind of political conflict that these different strategies are likely to provoke should be kept in mind. Given the institutional structure of the Community and the wording of the Treaties, it can generally be assumed that certain solutions will be preferred to others, not only because they are necessarily regarded as the most efficient, but also because they are more easily agreed upon than other policies. In other words, in the EU's institutional structure there is a structural bias towards favouring liberalisation and economic or 'soft' instruments.

The EU Treaty asks the Member States to develop a common transport policy and explicitly mentions the liberalisation of international inland transport. Accompanying measures are also mentioned as an option in Article 75 (1) d. However, this provision remains general and thus does not provide any narrowly defined need for action.[4] This distinction became important in the European Court of Justice (ECJ)'s inactivity verdict in 1985, when the Court on demand of the European Parliament stated that the Council had violated the Treaty by failing to establish a common transport policy within a reasonable period of time.[5] But the Court ruled that the Council's inactivity applied only in those cases where the Treaty was specific enough to define any concrete need for action, that is the liberalisation of transport as mentioned in Article 75 (1) a and b. The result was an imbalance in favour of liberalisation. The Council was directly called to liberalise market access and the ECJ explicitly decided that the liberalisation of services does not depend on the prior accomplishment of any other measures.

The Court thus put an end to a strategy some Member States had used to a sometimes excessive extent, which was agreeing to liberalisation only when parallel measures in related fields had been passed in the Council. Additional pressure arose from the theoretical possibility that in the case of continuing inactivity, the Treaty provisions could be put into practice through direct affect. But in the meantime, the Court left it up to the Council to decide upon the need for any further action, e.g. in the field of taxes, technical and social regulation, or infrastructure policies. Although the latter are often seen as necessary to ensure that a transport system works well, the ECJ's interpretation of the Community's legal framework does not exercise any concrete pressure to become active in these fields (Erdmenger, 1985).

Furthermore, diverging interests and variations in the policy approaches of the Member States exclude certain solutions from the beginning. Shifting goods from road to rail, for instance, has a completely different meaning in the Netherlands with the dominant position of road haulage, than it has in countries which have pursued a strategy of protecting national railways for decades. Furthermore, in addition to the fact that the basic economic ideology underlying European integration and the Single Market project is that of exploiting market forces, a country such as the United Kingdom, with its deeply rooted liberal tradition towards economic regulation, would hardly agree on policies that directly try to restrict market mechanisms.

Consequently, the Commission's publications, such as the recent Green and White Papers, although being generally favourable to shifts towards less environmentally damaging means of transport, always stress that freedom of choice for consumers has to be safeguarded. In other words, regulatory systems that, for instance, limit the available capacity in road transport – apart from having proved to be inefficient and to produce negative side-effects – could never be introduced at the Community level because of an opposing ideology dominating the supranational level, as well as the impossibility of agreeing on such measures in the Council of Ministers. On the contrary, the Commission repeatedly stresses that the abolition of rules restricting market access, except from certain qualification criteria that individual applicants for licences should meet, contributes to the most efficient utilisation of the existing capacity and thus represents a step towards making the transport systems less environmentally damaging.

Instead, the Commission puts great emphasis on economic strategies, namely pricing policies, which aim at improving the efficiency and environmental sustainability of the transport system by internalising infrastructure and external costs. This had already been the case with the 1992 Green and White Papers. Then, in December 1995 the Commission published another Green Paper entitled *Towards Fair and Efficient Pricing in Transport* (Commission, 1995a), which deals exclusively with the external costs of road transport. It points out deficiencies in the actual pricing systems and assumes that the existence of externalised costs favours road transport with respect to its competing modes. Hence, it indirectly calls for higher prices for road transport.

It is noteworthy how the Commission's approach emphasises the efficiency argument rather than that of environmental protection. Avoiding external costs is seen as identical with improving public wealth and ensuring fair competition. Thus, the Commission defines the internalisation of external costs as an economic and environmental win-win strategy rather than a matter of redistribution.[6] Independent of the question if this is really the case, it is interesting from a political point of view. Defining the problem in this way and linking it to a goal such as efficiency which, at least in theory, should gain unanimous approval, also increases the probability of approval among key policy-makers and can thus be regarded as a deliberate political strategy.

However, national policy-makers, as well as industry or transport associations, do not necessarily agree with such an approach. They might stress their short-term increases in costs rather than the assumed long-term efficiency gains for the whole transport and larger economic system. It is thus no surprise that the 1995 Green Paper has met the opposition of both road haulier organisations and transport users (i.e. industry) who heavily rely on road transport (UNICE, 1995; BDF, 1996). Consequently, the question arises whether the Community will be able to put into practice legislation concerning the internalisation of external costs after the Commission has put this issue on the agenda.

The conflicts which have arisen from the issue of liberalisation, and more specifically the harmonisation of legislation regarding the competitive position of transport enterprises, are likely to give way to some scepticism concerning the EU's capacity to introduce pricing systems which are effective enough to have an impact on the environmental performance of transport markets. The decision-making process regarding the common transport policy was deadlocked for a long time, because of diverging interests between Member States.[7] In addition, taxes turned out to be the issue where it was most difficult to reach agreement (cf. Schmitt, 1993; Mückenhausen, 1994). In this specific field, comprising vehicle and fuel taxes, tolls and road user charges, the delays have been the consequence of interest conflicts rooted in highly differing amounts of transit traffic and differences in the pre-existing national systems of infrastructure charging. Whereas some countries have applied tolls for motorways, others have relied on fuel taxes and vehicle excise duties as the main instrument for financing their infrastructure.

The Transport Council of Ministers in June 1993 finally agreed on the introduction of a new instrument of charging for infrastructure costs – the 'Eurovignette'. This imposes road user charges on an annual, weekly or daily basis in the Benelux countries, Denmark and Germany, who will be joined by Sweden in 1997 (Council Directive 93/98/EEC).[8] The introduction of road user charges in those countries which did not have such a system before can be interpreted in two ways. On the one hand, the level of taxation was set on a rather low level, with a ceiling of 1250 ECU annually. The same holds true for the minimum level for vehicle taxes which was introduced at the same time. Given the aforementioned discussion on external costs of transport, this can be regarded as disappointing from an environmental point of view. On the other hand, countries that did not apply road user charges before now have a new instrument at their disposition. The 1993 compromise might be a first step towards a system of charging transport that is truly based on the principle of territoriality, in other words, which allows to charge every single journey rather than using the crude instrument of annual, monthly, or daily charges, as it is the case with the Eurovignette.[9]

For the moment much will depend on the actual levels of taxation that the Council of Ministers will be able to agree upon in the future.[10] The question is whether the 'Eurovignette' can be developed further in the direction of a road-pricing mechanism, which actually gives sufficient incentives for the efficient use of transport infrastructures, leads to the stepwise internalisation of the external costs of transport, and influences the demand for transport. The Commission has now proposed a maximum level of 2500 ECU annually for the Eurovignette in the case of older vehicles, while trucks meeting the EURO I or EURO II emission norms would benefit from a price reduction. In addition, the Commission wants to introduce the possibility of raising extra charges for highly sensitive or congested areas (Commission, 1996b). This proposal has met with the opposition of the industrial and road hauliers' lobbies,[11] and the Council of Ministers, which at its meeting on 3 and 4 October 1996 could not agree on a new Directive. Instead, it asked the COREPER to continue deliberation on the issue (EC Council press release).

The Germans, for instance, support tripling the costs of the Eurovignette; but apart from the discussion about covering infrastructure and external costs, this also results from a general fiscal interest in charging foreign road users. The Dutch, on the other hand, have always preferred raising fuel taxes in order to contribute to the internalisation of external costs. This alternative would also be preferable from an environmental point of view, as fuel prices are better related to the real use of infrastructure and the actual amount of operational pollution that is produced than annual charges. But from the German point of view there is not sufficient guarantee that fuel taxes will benefit the German budget, as lorries could buy their fuel in other countries. In addition, increasing fuel taxes is also politically more sensitive, as it would raise the costs of private car use. A sum of 2500 ECU annually per truck cannot be regarded an excessive financial burden. However, as voting in the Council will be subject to the unanimity rule (as foreseen by Article 99 of the EC Treaty), negotiations are certain to remain difficult.

A further problem connected to raising the prices for road transport is the availability of alternative transport capacities. There might be unused capacities in rail transport or inland waterway shipping, but it has to be ensured that the alternative offer is of a similar quality, i.e. in terms of flexibility, reliability and speed. Otherwise, new pricing instruments would result in higher costs for consumers and higher tax revenues, but they will only have limited effects upon the organisation of production processes and the choice of certain means of transport. The declining importance of bulk goods and new forms of organising production processes, such as outsourcing and just-in-time deliveries, led to an increase in transport frequencies and fewer goods shifted in a single transport operation. This development has not only resulted in an overall increase of transport but also in the diminished importance of those goods which are most suitable to rail and ship transport. In addition, those goods which are typically transported by road haulage, and whose importance on transport markets is still increasing, are also those with the lowest price elasticity and the lowest transport costs in relation to the product's value (cf. Ginter and Schmutzler, 1996; Oum, Waters and Yong, 1992).

Consequently, pricing policies alone may not be sufficient to shift considerable amounts of goods between transport modes. Further steps need to be taken to make rail transport and inland waterway shipping more attractive and more responsive to the qualitative evolution of transport demand. Investment in railways and in the infrastructure of combined transport is needed, as well as an increase in the flexibility of the environmentally friendly modes of transport. Especially in international transport the competitive advantages of road haulage are partly due to coordination problems between the national railway companies, lack of technical harmonisation, etc.

The Commission published a White Paper on the Community's policy regarding the railways in July 1996 (Commission, 1996a). This paper tries to advance developments that started with the adoption of the Directive 440 in 1991. One of the most interesting aspects of this approach is that, at least according to the Commission's view, to make the railways more competitive, they should be no longer protected from competition in transport markets. On the contrary, they should be made more sensitive to market forces and pressures, for instance through competition between different railway companies on the same territory. Thus, with regard to the topic of deregulation (or liberalisation) and the environment, the railways might be an example of how the creation of markets does not only result in an enormous need to issue new kinds of regulation, but might indeed push reform processes that are absolutely necessary if there is to be a chance for reversing some of the negative trends that have been witnessed during the last decades.

However, this is an important point of departure from many of the Member States' classical approaches towards railway policy. Hence, independently of the accuracy of the Commission's argumentation, its proposals meet resistance, because the related reforms presuppose far-reaching learning processes and institutional adaptation at the domestic level. Such factors help to explain why some of the Member States have been reluctant to implement the Directive 91/440/EEC and why they are not very keen on exploring how the Commission's recent proposals in the White Paper could be put into practice (Agence Europe 4.10.1996).

CONCLUSION

Due to a variety of intervening factors it is extremely difficult to assess the exact environmental impact of economic regulation or deregulation in the transport sector. However, some tentative conclusions about the way the problem is dealt with politically within the EU framework can be drawn. Practical experience with existing regulatory systems has shown that quantitative restrictions for market access and price-setting by administrative bodies have not been able to guarantee a sensible allocation of goods, either in economic or in environmental terms. Thus, policies which have aimed at abolishing this kind of regulation in the transport sector in theory should be environmentally neutral. This however depends on certain preconditions such as, for instance, a sensible distribution of costs. According to many studies, these preconditions are not yet met in the transport sector.

The EU Commission is currently searching for strategies to make environmentally friendly modes of transport more competitive, without interfering directly with the transport users' freedom of choice. It strongly advocates the internalisation of the full costs of transport into prices and, subsequently, an extended use of fiscal instruments. It should be kept in mind that independently of its environmental effectiveness, this strategy is also the consequence of the specific institutional setting of the European Community. The utilisation of the efficiency argument in conjunction with environmental protection presents an opportunity for overcoming the resistance of opposing interests, as it is congruent with the Community's implicit economic ideology.

However, although important efforts towards making transport more compatible with the concept of sustainable development have been made on the programmatic level, there are good reasons for remaining sceptical about the Member States' willingness to agree on the Commission's new approach beyond rhetorical approval. The EU's intrinsic need to accommodate a wide variety of national and sectoral interests will continue to make it difficult to reach consensus about the rising costs of road transport or to agree on strategies for reforming the railways.

ACKNOWLEDGEMENTS

The author is grateful to Ute Collier and Dieter Kerwer for helpful comments on an earlier version of this paper.

NOTES

1 It is not against European law to keep certain national restrictions, provided that these rules are only applied to national carriers. However, as in the meantime national markets will be opened step-by-step to foreign operators, it would not be sensible for a government to impose restrictions on its own national carriers which put them at a disadvantage in competition with foreign hauliers.

2 An empirical study by Baum and Sarikaya (1995: 116ff.) gives the following figures referring to average load factors on German road haulage markets: ca. 58 per cent with regard to weight and 64 per cent with regard to volume in hire and reward haulage, whereas the figures in own account transport are 47 per cent with regard to weight and 49 per cent with regard to volume utilisation.

3 As can be seen by the reactions to the Commission's 1992 Green Paper (Commission, 1992b: no. 158).

4 The same is true for Article 130r (2) 2, which states that environmental protection requirements must be integrated in other EC policies.

5 Case 13/83, decision of 22.5.1985, OJ C 144, page 4; cf. Article 175 EC Treaty.

6 In addition to environmental reasoning, it is also sustained that transport volume management is becoming necessary because certain infrastructures cannot be extended any further (cf. Commission, 1992: no. 162).

7 On one side were the large and centrally located countries such as Germany, Italy and France, which tried to protect their national railways and the resident hauliers as far as possible from international competition. They generally used the argument that liberalisation could only be accomplished if parallel agreement on harmonisation in the fields of technical and social regulation, and taxes was reached. On the other side were the Benelux countries and those who entered the Community at a later point in time, who due to their geographical location, regulative traditions and expectations to gain from open markets (in terms of finding new fields of activity for their highly competitive national haulier fleets), gave absolute priority to the goal of liberalisation (Erdmenger, 1983).

8 The compromise was a package deal that, apart from the introduction of minimum levels for vehicle taxes and the introduction of new road user charges, also contained the stepwise liberalisation of cabotage by 1998 (Council regulation 3118/93/EEC).

9 Annual charges are not considered the best instrument to achieve an environmentally sensible use of transport capacities because once an annual charge has been paid, there is no further incentive to reduce transport activities. On the contrary, the more a vehicle gets used, the lower is the impact of the extra charge per kilometre.

10 According to the original Directive 93/89/EEC, the agreed minimum level of taxes was to be valid until the end of 1997 and open to revision afterwards. Road user charges were scheduled to be revised one year earlier, and every second year afterwards. Due to the intervention of the European Parliament, complaining that it had not been consulted properly, the European Court of Justice in 1995 declared the 'Eurovignette' Directive invalid (judgment of 5.7.1995 in case C-21/94, OJ C 229, p. 8). Thus the Directive has to pass the EU legislation process again in 1996.

11 Personal interviews with a member of Commissioner Kinnock's cabinet and the Brussels representative of the German road-haulier association BDF (17/19 April 1996).

REFERENCES

Barde, J.-P. and Button, K. (eds) (1990) *Transport Policy and the Environment*, London: Earthscan.

Baum, H. with Gierse, M. and Massmann, C. (1990) *Verkehrswachstum und Deregulierung in ihren Auswirkungen auf Strassenbelastung, Verkehrsicherheit und Umwelt*, Frankfurt a.M.: Verband der Automobilindustrie (VDA).

Baum, H. and Sarikaya, M.H. (1995) 'Umweltsteuern als Instrument zur Verringerung von Schadstoffemissionen im Straßengüterverkehr', *Zeitschrift für Verkehrswissenschaft* 66 (2), pp. 113–164.

BDF, Bundesverband des Deutschen Güterfernverkehrs (1994) *Jahresbericht 1993/94*, Frankfurt a.M.: BDF.

—— (1995a) *Jahresbericht 1994/95*, Frankfurt a.M.: BDF.

—— (1995b) *Flexible Kapazitätssteuerung im Straßengüterverkehr*, Frankfurt a.M.: BDF.

—— (1996) 'Grünbuch der EU über die Internalisierung der externen Kosten', unpublished paper (Rundschreiben Nr. A 1/96 vom 2.1.1996), Frankfurt a.M.: BDF.

Button, K. (1991) 'Regulatory reform', in K. Button and D. Pitfield (eds) *Transport Deregulation*, London: Macmillan.

—— (1992) *Market and Government Failures in Environmental Management*, Paris: OECD.

Chow, G. (1991) 'US and Canadian trucking policy', in K. Button and D. Pitfield (eds) *Transport Deregulation*, London: Macmillan.

Commission of the European Communities (1992a) *The Impact of Transport on the Environment*, COM (92) 46 final, Luxembourg: Office for Official Publications of the European Communities.

—— (1992b) *The Future Development of the Common Transport Policy*, COM (92) 494 final, Luxembourg: Office for Official Publications of the European Communities.

—— (1995a) *Towards Fair and Efficient Pricing in Transport*, COM (95) 691, Luxembourg: Office for Official Publications of the European Communities.

—— (1995b) *Progress Report from the Commission on the Implementation of the European Community Programme of Policy and Action in Relation to the Environment and Sustainable Development*, COM (95) 624 final, Luxembourg: Office for Official Publications of the European Communities.

—— (1996a) *A Strategy for Revitalising the Community's Railways*, COM (96) 421 final, Luxembourg: Office for Official Publications of the European Communities.

—— (1996b) *Proposal for a Directive replacing Directive 93/89/EEC on Heavy Vehicle Taxes, Road User Charges, and Tolls*, COM (96) 331 final, Luxembourg: Office for Official Publications of the European Communities.

Crawford, I. and Smith, S. (1995) 'Fiscal instruments for air pollution abatement in road transport', *Journal of Transport Economics and Policy* 39 (1), pp. 33–51.

Deregulierungskommission. Unabhängige Expertenkommission zum Abbau marktwidriger Regulierungen (1991) *Marktöffnung und Wettbewerb*, Stuttgart: C.E. Poeschel.

DIW, Deutsches Institut für Wirtschaftsforschung (1994) *Verminderung der Luft- und Lärmbelastungen im Güterfernverkehr 2010*, Berlin: DIW.

ECMT, European Conference of Ministers of Transport (ed.) (1994) *Internalising the Social Cost of Transport*, Paris: OECD.

Enquete-Kommission 'Schutz der Erdatmosphäre' des Deutschen Bundestages (1994) *Mobilität und Klima*, Bonn: Economica.

Erdmenger, J. (1983) *The European Community Transport Policy*, Aldershot: Gower.
—— (1985) 'Die EG-Verkehrspolitik vor Gericht', *Europarecht* 20 (4), pp. 375–392.
Ginter, D. and Schmutzler, A. (1996) 'Die Aufteilung des Güterverkehrs auf Bahn, LKW und Schiff – eine dynamische Analyse', *Zeitschrift für Verkehrswissenschaft* 67 (1), pp. 49–67.
Hamm, W. (1989) *Deregulierung im Verkehr als politische Aufgabe*, München: Minerva.
Hey, C. (1994) *Umweltpolitik in Europa*, München: Beck.
Mückenhausen, P. (1994) 'Die Harmonisierung der Abgaben auf den Straßengüterverkehr in der Europäischen Gemeinschaft', *Europäische Zeitschrift für Wirtschaftsrecht* 5 (17), pp. 519–523.
Oum, T.H., Waters, W.G.I. and Yong, J.-S. (1992) 'Concepts of price elasticities of transport demand and recent empirical estimates', *Journal of Transport Economics and Policy* 26 (2), pp. 139–154.
Schmitt, V. (1993) 'Die Harmonisierung der Wettbewerbsbedingungen in der EG-Binnenverkehrspolitik', *Europäische Zeitschrift für Wirtschaftsrecht* 4 (10), pp. 305–311.
UNICE, Union des Confédérations de l'Industrie et des Employeurs d'Europe (1995) *UNICE's Views on Recent Work concerning Internalisation of the External Costs of Transport*, Brussels: UNICE.

8

MARKET AND REGULATORY FAILURE IN THE WATER SECTOR

Simon Cowan

INTRODUCTION

Environmental policy for the water sector is an area of key public concern, and legislation and implementation are developing, albeit slowly, within Europe. The Fifth Community Action Programme on the Environment, which was adopted in 1993, stressed the need to complement legislation with the use of market-based instruments including environmental charges, civil liability schemes and voluntary agreements. For the management of water resources the targets for 2000 were prevention of over-exploitation of ground and surface water for drinking and industrial purposes, prevention of pollution of groundwater from diffuse sources and better ecological quality of surface and marine water.

This chapter examines the role of alternative instruments for water pollution from an economic perspective. Water pollution has characteristics that differ from other types of pollution, and these characteristics are important in determining the extent to which market-based instruments can be used. The chapter begins its analysis with a discussion of the characteristics of water pollution. Then, the possible instruments that can be applied to the case of water pollution are considered, namely command-and-control regulation, the use of liability rules, economic instruments and voluntary agreements. The chapter then describes and assesses some of the current policies towards water pollution within Europe and finishes with some concluding remarks.

WATER POLLUTION

The water supply process involves abstraction, treatment and distribution. Raw water is abstracted from surface sources such as rivers, lakes and reservoirs, and from underground sources. In Europe as a whole about 70

131

per cent of abstraction for all purposes comes from surface sources, although there is considerable variation between countries (Stanners and Bourdeau, 1995, p. 62). Both surface and ground sources are usually replenished by winter rainfall. Raw water is treated to ensure that it can be safely drunk. Treatment typically involves filtration to remove suspended matter, disinfection with chlorine to kill harmful bacteria and pH correction to minimise corrosion in the distribution network.

Domestic households use very little of the water they consume for drinking. Most is used for washing and flushing lavatories. Industrial customers use water in their production processes. One large source of demand comes from power stations that need raw water as a coolant. Most publicly supplied water returns as sewage to the sewerage system. Industrial waste water can vary in strength and nature and is usually priced accordingly. Industrial customers can reduce their costs by pre-treating effluent before it enters the sewer or by bypassing the sewerage company and doing all the treatment on site. Sewage is pumped to treatment works where solids and harmful bacteria are removed by sedimentation. Sewage sludge is dumped in landfill sites or at sea, used as fertiliser on farmland, or incinerated. Effluent, the liquid product of treatment works, is returned to rivers or estuaries. Raw sewage could also be pumped out to sea without treatment.

Some pollutants can be broken down by bacteria. Rivers, lakes and seas can purify themselves if they have sufficient dissolved oxygen to support these bacteria. The amount of dissolved oxygen in a river will fall if the flow of the river falls and if the volume or strength of the effluent discharged into it increases. A measure of the oxygen demand placed on a stream by any particular volume of effluent is biochemical oxygen demand (BOD). Thermal pollution, caused by a discharge of warm water from a power station, also lowers the dissolved oxygen content of a river. Other pollutants, however, accumulate and are not naturally broken down and are thus particularly awkward to eradicate. Examples are aluminium, and heavy metals such as cadmium, mercury and lead. Water pollution is usually a local or regional phenomenon, unlike the global problem of carbon emissions and the international problem of the ozone layer.

The main point sources of water pollution are sewage treatment works and industrial plants. Sewage treatment works dispose of effluent and sludge, and the latter can be a pollutant of the water environment if it is dumped at sea or in landfill sites. If sludge is incinerated then there can be consequences for air pollution. Nonpoint-source pollution derived, for example, from agricultural activity, is also important. Pesticides and nitrates can eventually reach aquifers, rivers and lakes. Since the sources of such pollution cannot be traced only indirect methods of control are feasible, such as taxing the purchase of fertilisers or pesticides, or restricting their use in sensitive areas.

There is a close relationship between the quality of water that is consumed and the amount of pollution in the water environment, since extra pollution increases the costs of meeting a given standard of water purity. Water quality has many dimensions. Taste, odour and colour are all readily discernible by consumers. Other aspects of quality are less observable before consumption takes place, such as concentrations of lead, copper and zinc, and the presence of other harmful substances. The dimensions of water quality that are most important for health are unfortunately those that customers are unable to inspect before consumption.

INSTRUMENTS FOR WATER POLLUTION PROBLEMS

In this section I assess the different instruments that can be used to control water pollution. First I discuss the reasons for controlling pollution. A key concept is that of *market failure*. This exists if unrestricted market forces generate an allocation of resources that is inefficient, in the sense that it would be theoretically possible to make at least one economic agent better off without harming other agents. Under well-known conditions free markets that are characterised by perfect competition generate *efficient* outcomes.[1] The problem, however, is that such ideal conditions often do not hold. Environmental pollution is an example of a ubiquitous type of market failure, that associated with the presence of *externalities*.[2] An example in the water context is the case where a sewage treatment plant is located upstream of an angling club, and the plant does not take account of the loss of amenity to the anglers caused by the effluent. The polluter will emit effluent until the point at which the marginal net benefit to him of further pollution is zero. Since this calculation does not include the costs to others the outcome is socially inefficient and there is a potential role for public policy.

Having established that there is at least a *prima facie* case for government intervention when there is market failure, a caveat should be noted. Governments or their agencies can fail as well as markets. Agencies can be corrupt or inefficient in setting or enforcing standards. An agency is unlikely to have the same information as polluters about the technologies and costs involved in pollution control and this asymmetry of information could lead to regulatory capture, when the agency identifies with the interests of the regulated firm. Alternatively the government or the agency might simply adopt inefficient policies for various non-economic reasons. Cropper and Oates (1992, p. 675) note that 'the cornerstones of federal environmental policy in the United States, the Amendments to the Clean Air Act in 1970 and to the Clean Water Act in 1972, *explicitly* prohibited the weighing of benefits against costs in the setting of environmental standards'. Such prohibitions make the achievement of economically efficient standards, which by definition must balance costs against benefits, difficult to attain.

133

In Europe the most common form of regulation is of the command-and-control type. With this type of regulation governments, or their agencies, determine acceptable standards for each polluting source. Such standards might take the form of quantitative restrictions on the volume or strength of emissions a source can make or they might prescribe particular technology, such as secondary treatment of sewage, that must be installed. The main problem from an economic point of view is that if standards are set that require uniformity, such as equal percentage reductions in emissions from some baseline level, then it is very unlikely that the cost of achieving any given aggregate level of abatement will be minimised.

To focus on the abatement costs issue suppose that pollution damage is not location-specific, which is true for some types of air pollution. Thus the amount of damage is determined by the total quantity of pollution, and if increases in emissions from one source are balanced by reductions elsewhere then damage is unaffected. In general a policy requiring uniform abatement is inefficient. Some sources will find an extra unit of abatement to be very expensive, while others will be able to abate much more cheaply. Abatement costs will be saved by encouraging the high-cost sources to do less abatement and the low-cost sources to do more with no change in the damage done. Empirical studies of air pollution control suggest that the extra cost of abatement caused by uniform command-and-control standards can be very significant. For example, Tietenberg (1991, Table 4.1) quotes studies from the United States where the ratio of command-and-control abatement costs to theoretical least costs varies from 1.07 to 22.00.

Liability rules

Liability rules are attracting increasing attention in Europe following the publication of the Commission's Green Paper *Remedying Environmental Damage* in 1993. The idea is that property rights are allocated to potential victims of pollution, and they can recover damages under certain circumstances.[3] The externality will then be internalised as the polluter is forced to take account of the damage caused by his pollution. If liability rules substitute for more explicit forms of regulation the process can be thought of as privatisation of the public sector's role of setting, monitoring and enforcing given standards.

This might be preferable to public regulation because action is only taken where damage has occurred rather than in all states of the world, and because those who are harmed by pollution might be better at seeking recompense than public authorities. The main rules are *strict liability*, which requires the polluter to pay damages equal to the costs that victims suffer whether or not it was the polluter's fault, and a *negligence* rule under which the polluter is only liable for damages if he did not take an appropriate level of care.[4] In theory at least a negligence rule is more efficient than strict liability when the probability of a pollution incident partly depends on actions that the *victim* can take (see Segerson, 1995). Under strict liability the victim is fully compensated for all damages and he would then have insufficient incentive to take the appropriate level of care.

In practice, however, a negligence rule might be more expensive to operate than strict liability, as the lawyers need to determine not only who caused a pollution incident but also whether the amount of care taken was adequate. The extra complication of a negligence rule argues against it, though the relative costs of implementation should be determined empirically. In some cases it might be relatively straightforward to determine whether there was negligence such as in a case where optimal care is taken if and only if a particular piece of capital equipment is installed.

In the water context there are some problems with the application of liability rules. The existence of nonpoint sources of pollution means that it might be impossible in practice to attribute amounts of pollution to particular sources.[5] Even when the sources of pollution can be traced there is a further problem of valuing the damage caused, especially when the victims are individuals rather than firms. Menell (1991, p. 94) argues that 'it is extremely costly, if not scientifically impossible, to link a particular injury or disease to a particular environmental cause'. In the context of long-term liabilities, liable parties may go bankrupt to avoid paying for the damage.

Most water pollution takes the form of continuous emissions rather than occasional, accidental spills, and the litigation costs of dealing with these under a strict liability scheme would be enormous. The costs would be much smaller if a workable negligence rule is used, because only those harms caused by insufficient care, rather than all harms, would create liabilities. It makes more sense to apply command-and-control regulation or an economic instrument such as a tax that continuously provides the polluter with the right incentives without the transaction costs involved in litigation. A further problem is that water pollution usually affects a large number of victims or potential victims. The likelihood that these victims will be able to act together to claim the correct compensation is small given the free-rider problems and the transaction costs involved, although it should be noted that such agreement is more likely in the context of water pollution than air pollution, where some of the victims are not even born yet.

Even if there are small numbers of victims it is likely that bargaining and legal costs would be substantial. Porter and van der Linde (1995, pp. 115–116) quote a study that estimates that '88 per cent of the money spent by insurers between 1986 and 1989 on Superfund claims went to pay for legal and administrative costs, while only 12 per cent was used for actual site cleanups'. Strict liability also gives victims perfect insurance, which leaves them over-insured compared to what they would have bought privately (Shapiro, 1991). Finally liability rules do not work well when a significant part of the environment is unowned (Grant, 1996). Although economists have devised the contingent valuation technique to estimate the 'non-use' value of environmental assets, there is considerable controversy about its use (for opposing views see Portney, 1994; Hanemann, 1994; and Diamond and Hausman, 1994).

Although I have argued that liability rules would make a poor substitute for existing forms of regulation of standard water pollution it can be argued that they can play a useful role in preventing *accidental* spills of dangerous substances, and there are reasons for believing that their application is easier in the context of water pollution than air pollution. Ball and Bell (1995) list four reasons why water pollution cases under common law have proved easier to bring in the UK than air pollution cases: causation is easier to show in the water context; many rural landowners are rich enough to afford the costs of legal action; there exist several campaigning organisations that effectively overcome the free-rider problem mentioned earlier; and the acquisition of evidence is more straightforward because of the existence of public pollution registers. The UK also uses *criminal* liability for serious breaches of discharge consents.

Economic instruments

The third class of instruments for environmental regulation are the economic instruments. The two main types are taxes and tradable permits. The economic case for these is now well known so the discussion here will be brief. Governments have been slower to introduce them than might have been expected given their theoretical attractions. One argument for a pollution tax is that this causes effective internalisation of the externality by raising the cost to the polluter of his activity. If the tax is set equal to the marginal damage cost then the polluter has the correct incentive for the choice of pollution level. Even if the tax is not equal to the marginal damage the fact that the tax is the same for all pollution sources means that the cost of achieving the given level of abatement is minimised. A pollution tax thus provides an automatic mechanism to minimise the cost of achieving any given level of pollution abatement.[6]

It should be noted that the theoretical case for a pollution tax does *not* depend on the revenue raised being used to compensate victims (indeed victim compensation can be subject to some of the objections to the strict liability rule given above). In practice, however, there will be strong pressure to recycle the revenue generated by a pollution tax either to compensate victims or to subsidise further pollution-control measures.

A tradable permit system ensures that the total cost of abatement is minimised because marginal incentives are provided by the market price of a permit which is the same for all pollution sources. Such a scheme works in a similar way to a tax, though there are differences between taxes and permits when there is uncertainty.[7] In the United States permit schemes are now common for air pollution and there has been some use of them for water pollution. The main problem with permit schemes for water is that pollution is a local phenomenon so trading would need to be restricted, for example to a particular river basin or lake. This restriction on trading is likely to lead to very thin markets for the permits, with consequent price distortions. A permit scheme was introduced by the State of Wisconsin to control BOD in the 1980s in the Fox River, but in the first six years of operation there was only one trade (see Hahn, 1989; and Tietenberg, 1992, p. 496). Pollution control instruments can be combined, and this can be preferable to exclusive use of one instrument (Roberts and Spence, 1976). As we shall see in the next section in Europe it has been common to use several instruments together (usually command-and-control plus pollution charges).

In general economic instruments and liability rules are seen as substitutes and not complements. In cases where a liability rule works well it is superfluous to have a pollution tax at the same time and *vice versa*. The choice between them should depend on the relative costs of implementation. Economic instruments are really a form of regulation that requires a government agency to set taxes or quantity of pollution, to allocate permits and to monitor emissions. Such an agency requires resources. In theory there is no need for any government agency under a system of liability, but there will be litigation costs. The problem with the argument that there should be either regulation or liability but not both, however, is that it is predicated on the assumption that each scheme works efficiently on its own. If this is the case then it is obvious that to combine liability and regulation is to overdo it. In practice, of course, neither scheme is perfect. Shavell (1984) and Kolstad *et al.* (1990) present arguments for combining regulation and liability when neither scheme is fully efficient on its own, and argue that increasing the role for liability entails that existing regulation should be relaxed.

Voluntary agreements in the environmental context pose some conceptual problems for economists. If economic agents optimise then it is difficult to understand why they will want voluntarily to cut back their pollution, unless it is under threat of more draconian measures. The motives for voluntary agreements can be more sinister than might at first be apparent. A voluntary agreement to cut pollution will usually be specified in quantitative form, e.g. emissions cannot rise at each source above a historically determined level. Such an agreement might be used to keep new firms out of an industry. Indeed Article 3 of the European Commission's action plan adopted in January 1996 ('Towards Sustainability') stated that 'special attention will be given to . . . voluntary agreements in the field of the environment *in conformity with competition rules*' [my italics], so at least this concern has been noted.[8]

The danger with putting too much reliance on voluntary agreements is that necessary regulatory action might be put off in the vain hope that industry will voluntarily control emissions. Having said that, there are some advantages to voluntary agreements. First, creating and enforcing legislation is expensive, and voluntary agreements will save some of these costs. Second, it can be argued that there are many free lunches waiting to be taken. Porter and van der Linde (1995) argue that environmental regulations, if designed with sufficient flexibility and certainty, can foster innovation and induce firms to act efficiently when they previously were not efficient.[9] If this is the case then a voluntary agreement on pollution control might likewise benefit not only the environment but also the polluters.

CURRENT ENVIRONMENTAL REGULATION OF WATER IN EUROPE

The main existing European Directives on water are:

- 76/160 on Bathing Water;
- 76/464 on Dangerous Substance Discharges;
- 80/778 on Water for Human Consumption;
- 91/271 on Urban Waste Water Treatment.

The Bathing Water Directive lays down 19 parameters with which all bathing waters (both fresh and marine) must comply. The Dangerous Substances Directive specifies two lists of substances to be eliminated or reduced. The Water for Human Consumption Directive is concerned with consumer protection rather than environmental regulation. It specifies 62 parameters relating to the quality of water provided for human consumption or for the purposes of food manufacturing. The Urban Waste Water Treatment Directive lays down minimum standards for the treatment of urban waste waters. Essentially it requires that secondary treatment should be usual for domestic wastes, and under it the dumping of sewage sludge at sea must cease by the end of 1998.

Under the principle of subsidiarity each Member State can choose how to implement the Directives. The record of Member States in implementation and enforcement of water Directives has been particularly poor. The standard procedure has been to use command-and-control methods rather than to rely on civil liability, economic instruments or voluntary agreements. In England and Wales the main anti-pollution agency is the new Environment Agency (EA), which started operation in April 1996 and is an amalgamation of the National Rivers Authority (NRA) and Her Majesty's Inspectorate of Pollution (HMIP). There is a separate agency for Scotland. The Environment Agency has the duty under the Environment Act 1995 to take into account both the benefits and the costs of particular environmental protection measures, although as is common in UK legislation the definitions of these terms are left unclear. In spite of general government interest in economic instruments the UK system remains one of command

and control. The EA sets water quality objectives for inland, coastal, estuarial and ground waters and polluters are required to obtain a 'discharge consent' to be entitled to discharge into these waters. A discharge consent specifies the maximum volume and strength of effluent that can be put into a water course. No reallocation of consents is allowed. The system is backed by both the criminal law and some well-developed common law principles.

The NRA and its successor the EA have set annual charges for discharge consents since 1991. Charges are the same throughout the country, relate to what is consented rather than to the actual discharge, and are designed in the long run to recover the administrative costs that the EA incurs for running the consent system. Smith (1995, p. 32) notes that 'the structure of the tariff is intended to reflect the costs that each discharger imposes in terms of monitoring and compliance work, rather than the pollution damage of the effluent. Thus the tariff rises with the size of the discharge, but less than proportionately with volume, since monitoring costs do not rise proportionately.' Although such charges are not designed to influence abatement choices they have the potential to do so, especially if they are raised. It should be noted that the EA charges are an addition to the system of consents and not a substitute for it. Helm (1993) argues for a stronger version of the current system that retains the existing consents and sets higher, incentive-based pollution charges. If abatement turns out to be very expensive relative to the tax level then the quantitative constraint of the consent will bind, and pollution will be at the level implied by the consent. But if abatement turns out to be relatively cheaper this combined scheme would provide an incentive for the polluter to abate beyond the level required by the consent to save on tax payments.

A peculiar feature of the UK water industry is that more than 90 per cent of domestic customers do not have water meters. Such customers do not face any marginal incentives to constrain water use or discharges to sewers. Without meters no price mechanism designed to give domestic households the proper environmental signals can be used. It can be argued, however, that pollution charges on sewage effluent will give the water and sewerage companies stronger incentives to install water meters to measure water consumption and, indirectly, household discharges to sewers.

Has privatisation of the water and sewerage companies in England and Wales been a success in terms of environmental protection? There has been criticism in the media of 'excessive' profits, over-generous remuneration packages for top management and widespread rationing during the drought of 1995. Nevertheless there have been considerable benefits. The regime is more explicit than the previous one, in which publicly owned water authorities were responsible for environmental regulation in addition to the core functions of water supply and waste-water treatment and disposal. The environment was not well protected during the era of all-purpose water authorities (Kinnersley, 1988). Under the current regime prosecution of polluters who breach their consents has proved much easier, essentially because the water companies are no longer regulating themselves. There is also an argument for saying that privatisation in 1989 has meant that the capital investment necessary to meet the EU requirements was undertaken more quickly than it would have been if the water companies had remained in public ownership. The government privatised the investment programme because it was unwilling to invest the money itself. Indeed there has been a boom in investment by the water companies since 1989, with most of the money being spent on improvements to sewage treatment and disposal.

In other European countries there are a variety of taxes on direct discharges (in addition to the usual charges for 'indirect discharges' to public sewers) that are closer than the UK charges to the economist's ideal of a tax that has incentive effects. See Commission of the European Communities (1995) for a comprehensive description of the schemes in place in 1994. The most well-known examples are France, Germany and the Netherlands, but other countries such as Belgium, Portugal and Spain also have water effluent charges (see OECD, 1994, Table 3.8). It should be noted that in all these cases the taxes are in addition to existing direct regulations. France has levied charges since 1968. Large dischargers (defined as those that discharge pollution loads in excess of 200 'inhabitant equivalents') are charged on the basis of loads estimated by the Agence de l'Eau, the regional water agency, although either party can ask for charges to be based on direct measurement rather than estimates. The main objective of the charges is to provide revenue for investment in sewage treatment and other water pollution control measures. The effects of the charges on pollution incentives are unclear, though studies have suggested that the levels are not high enough to promote abatement.

In Germany the water pollution charges are administered by the *Länder*, which use the revenue to cover administrative costs and for waste water treatment. Charges are uniform across the country, and have risen significantly in real terms since their introduction in 1981 (Smith, 1995, Table 4.2). The charge can be reduced if the source demonstrates compliance with a new standard before the latter becomes binding. It appears that the charges have had significant incentive effects (though it is difficult to disentangle the effects of the charges from those of the associated direct regulations). In particular there was a large increase in abatement investment in the run up to the introduction of the charges.

In the Netherlands the charges are set to finance sewage treatment costs. Households and small firms pay flat-rate charges, firms that have an intermediate pollution load can choose whether to pay on the basis of predicted or actual loads, while large sources pay according to the quantity and concentration of actual emissions. Charges have risen over time, and they differ between regions. There is some evidence of an incentive effect for large sources that are metered, although again it is difficult to disentangle the effect of direct regulation from the effect of the effluent charges.

Note that water pollution charges are used *in combination* with existing command-and-control regulations of the consent type, rather than as substitutes for these regulations. The effectiveness of charges schemes on their own is necessarily uncertain given the lack of information about abatement technology and costs that governments have. By adding a charging scheme to existing regulations governments can raise revenue that can be used for further environmental improvements. Such recycling of revenue is also important for the acceptance of charging schemes by industry.

The simplest way to deal with nonpoint-source pollution is to tax the offending products. OECD (1995, pp. 36–7) summarises the position for OECD countries on environmental taxation of agricultural inputs. Austria has had a fertiliser levy since 1986. The Flemish Region in Belgium has levied a tax on surplus manure since 1991, and the federal government was due to introduce a tax on some pesticides in 1995. In Denmark pesticides are subject to a tax whose value depends on the quantity purchased. In the Netherlands there is a surplus manure tax, except for those farms producing small quantities. Sweden has levied a fertiliser charge since 1984.

A new Directive on the ecological quality of surface water has been proposed by the Commission. In accordance with the intentions expressed in the Fifth Environmental Action Programme the proposed new Directive has less emphasis on the command-and-control aspects of regulation than previous legislation. The Member States will have to monitor the overall ecological quality of their waters, define operational targets to maintain or improve ecological targets and establish integrated programmes to achieve these targets. Precise implementation, under the principle of subsidiarity, is up to the Member States, but decisions must be made in a transparent manner with full consultation. This proposal moves away from the earlier approach to water legislation which set limit values for a series of chemical or physical parameters, and adopts a more holistic, ecological approach. Changes to the existing Drinking Water and Bathing Water Directives are also proposed, and again the Commission has adopted a less *dirigiste* approach than in the original Directives. In particular the number of parameters monitored will be reduced and much better information for those affected will be required.

CONCLUSIONS

In this chapter I have discussed the environmental problems that are associated with water and the forms of regulation that have been adopted by the European Union and its Member States. The disappointing record of compliance with earlier EU Directives has led the Commission to recommend the use of a broader range of instruments than has been used hitherto. The analysis suggests that it is most sensible to use market-based instruments such as environmental charges in addition to, rather than instead of, quantitative regulations. This is the way that water-pollution charges have been applied in practice in continental Europe, although the precise incentive effects of many charging schemes are difficult to determine. Civil liability systems are not especially relevant for the bulk of water pollution but they might be of use to deter accidental spills of dangerous substances.

Privatisation of the water industry in the UK has had some benefits for the environment since it makes enforcement of legislation easier, makes the environmental objectives more transparent, and encourages investment spending that was not forthcoming when the companies were in public ownership. Private participation in water and sewerage services is being considered by many developing countries that are in fiscal difficulties, and other EU Member States might want to consider increasing the role of the private sector in the water and sewerage industry.

ACKNOWLEDGEMENTS

I would like to thank the conference participants, and Robin Mason and Ute Collier for very helpful comments on an earlier draft. They are not, of course, responsible for any errors.

NOTES

1 See any microeconomics text, such as Varian (1993).
2 See Baumol and Oates (1988, p. 17) for a comprehensive definition of an externality.
3 Liability rules can be thought of as a combination of the Coase Theorem (Coase, 1960), which claims that polluters and victims will bargain over the amount of pollution and achieve an efficient outcome provided transactions costs are negligible and property rights are clearly allocated to either party, with the Polluter Pays Principle that requires that the property rights are allocated to the victim.
4 For theoretical analyses of liability rules that might apply to the water sector see Wetzstein and Centner (1992) and the reply by Miceli and Segerson (1993).
5 For discussion of theoretical approaches to the problem of nonpoint-source pollution see Segerson (1988), Xepapadeas (1991, 1992) and Cabe and Herriges (1992).
6 A pollution tax also provides an incentive to invent new and cheaper methods of abatement in order to reduce the tax payments.
7 Weitzman (1974) and Baumol and Oates (1988) discuss the choice between taxes and permits.
8 We should also remember Adam Smith's famous warning: 'People of the same trade seldom meet together, even for merriment and diversion, but the conversation ends in a conspiracy against the public, or in some contrivance to raise prices' (*The Wealth of Nations*, 1776).
9 Palmer *et al.* (1995) argue against Porter and van der Linde's position.

REFERENCES

Ball, S. and Bell, S. (1995) *Environmental Law: The Law and Policy relating to the Protection of the Environment*, 3rd edition, Blackstone: London.

Baumol, W.J. and Oates, W.E. (1988) *The Theory of Environmental Policy*, 2nd edition, Cambridge: Cambridge University Press.

Cabe, R. and Herriges, J.A. (1992) 'The regulation of non-point-source pollution under imperfect and asymmetric information', *Journal of Environmental Economics and Management*, 22, pp. 134–146.

Coase, R.H. (1960) 'The problem of social cost', *Journal of Law and Economics*, 3, pp. 1–44.

Commission of the European Communities (1995) *Waste-Water Charging Schemes in the European Union*, Luxembourg: Office for Official Publications of the European Communities.

Cropper, M.L. and Oates, W.E. (1992) 'Environmental economics: a survey', *Journal of Economic Literature*, 30, pp. 675–740.

Diamond, P.A. and Hausman, J.A. (1994) 'Contingent valuation: is some number better than no number?', *Journal of Economic Perspectives*, 8, pp. 45–64.

Grant, M. (1996) 'Environmental liability', in Winter, G. (ed.) *European Environmental Law: A Comparative Perspective*, Aldershot: Dartmouth.

Hahn, R.W. (1989) 'Economic prescriptions for environmental problems: how the patient followed the doctor's orders', *Journal of Economic Perspectives*, 3(2), pp. 95–114.

Hanemann, W.M. (1994) 'Valuing the environment through contingent valuation', *Journal of Economic Perspectives*, 8, pp. 19–44.

Helm, D.R. (1993) 'Market mechanisms and the water environment: are they practicable?', in Gilland, A. (ed.) *Efficiency and Effectiveness in the Modern Water Business*, London: Centre for the Study of Regulated Industries, Public Finance Foundation.

Kinnersley, D. (1988) *Troubled Water*, London: Hilary Shipman.

Kolstad, C.D., Ulen, T.S. and Johnson, G.V. (1990) 'Ex post liability for harm vs. ex ante safety regulation: substitutes or complements?', *American Economic Review*, 80, pp. 888–901.

Menell, P.S. (1991) 'The limitations of legal institutions for addressing environmental risks', *Journal of Economic Perspectives*, 5, pp. 93–113.

Miceli, T.J. and Segerson, K. (1993) 'Regulating agricultural groundwater contamination: a comment', *Journal of Environmental Economics and Management*, 25, pp. 196–200.

OECD (1994) *Managing the Environment: The Role of Economic Instruments*, Paris: OECD.

OECD (1995) *Environmental Taxes in OECD Countries*, Paris: OECD.

Palmer, K., Oates, W. and Portney, P.R. (1995) 'Tightening environmental standards: the benefit-cost or the no-cost paradigm?', *Journal of Economic Perspectives*, 9, pp. 119–132.

Porter, M.E. and van der Linde, C. (1995) 'Toward a new conception of the environment–competitiveness relationship', *Journal of Economic Perspectives*, 9, pp. 97–118.

Portney, P.R. (1994) 'The contingent valuation debate: why economists should care', *Journal of Economic Perspectives*, 8, pp. 3–18.

Roberts, M.J. and Spence, M. (1976) 'Effluent charges and licenses under uncertainty', *Journal of Public Economics*, 5, pp. 193–208.

Segerson, K. (1988) 'Uncertainty and incentives for nonpoint pollution control', *Journal of Environmental Economics and Management*, 15, pp. 87–98.

Segerson, K. (1995) 'Liability and penalty structures in policy design', in D.W. Bromley (ed.) *The Handbook of Environmental Economics*, Oxford: Basil Blackwell.

Shapiro, C. (1991) 'Symposium on the economics of liability', *Journal of Economic Perspectives*, 5, pp. 3–10.

Shavell, S. (1984) 'A model of the optimal use of liability and safety regulation', *RAND Journal of Economics*, 15, pp. 271–280.

Smith, S. (1995) *'Green' Taxes and Charges: Policy and Practice in Britain and Germany*, London: Institute for Fiscal Studies.

Stanners, D. and Bourdeau, P. (eds) (1995) *Europe's Environment: The Dobris Assessment*, Copenhagen: European Environment Agency.

Tietenberg, T.H. (1991) 'Economic instruments for environmental regulation', in Helm, D. (ed.) *Economic Policy towards the Environment*, Oxford: Blackwell.

Tietenberg, T.H. (1992) *Environmental and Natural Resource Economics*, 3rd edition, New York: HarperCollins.

Varian, H.R. (1993) *Intermediate Microeconomics: A Modern Approach*, 3rd edition, New York and London: Norton.

Weitzman, M.L. (1974) 'Prices vs. quantities', *Review of Economic Studies*, 41, pp. 477–491.

Wetzstein, M.E. and Centner, T.J. (1992) 'Regulating agricultural contamination of groundwater through strict liability and negligence legislation', *Journal of Environmental Economics and Management*, 22, pp. 1–11.

Xepapadeas, A.P. (1991) 'Environmental policy under imperfect information: incentives and moral hazard', *Journal of Environmental Economics and Management*, 20, pp. 113–126.

Xepapadeas, A.P. (1992) 'Environmental policy design and dynamic nonpoint-source pollution', *Journal of Environmental Economics and Management*, 23, pp. 22–39.

Part III

INDUSTRY IN A DEREGULATORY CLIMATE

9

LARGE FIRMS, SMEs, ENVIRONMENTAL DEREGULATION AND COMPETITIVENESS

Wyn Grant

INTRODUCTION

Within the context of the impact of environmental deregulation on competitiveness, it is important to consider the different positions of large firms and small and medium-sized enterprises (SMEs). Following Brusco, Bertossi and Cottica (1996), it is suggested that firms are playing the game of competition on two different chessboards: the policy arena and the competitive arena.

> A firm's ability to influence the outcomes in the regulatory arena depends on its position in the competitive arenas; symmetrically, its competitive position depends, [in] the medium and long run, on its ability to secure advantages on the policy arena, so that the two chessboards interact with each other in a complex way.
>
> (Brusco, Bertossi and Cottica, 1996, p. 133)

This chapter suggests that, in general, the balance of advantage in both arenas is with the larger firm, although the difference is more marked in the policy arena than in the competitive arena. It is argued that large firms have greater opportunities to influence the agenda of environmental deregulation, but that, in general, the competitiveness of smaller firms is likely to be more adversely affected by existing and proposed regulations. Large firms are thus regulation makers, while SMEs are 'regulation-takers' (Lévêque, 1996a, p. 201). The chapter reviews the way in which ideas about environmental deregulation have developed in the United States, suggesting that they have been an influence on the debate in Europe which is also assessed. Because of the way in which large firms have a dominant role in the policy arena in relation to smaller firms, environmental deregulation could even occur in a way which undermines the market position of SMEs.

The effect of regulation on competitiveness is not, of course, the only relevant consideration in the formulation of environmental policy. Indeed, in the case of the Large Combustion Plants Directive, 'competitive considerations and the alignment of Member States behind their industry's interests was probably a more potent factor determining the policy outcome than was environmental ambition' (Ikwue and Skea, 1996, p. 93). The focus of this chapter means that the impact of environmental regulation and deregulation on competitiveness and market structures is given more emphasis than in other chapters. However, it is important to remember that, although any one SME may not be a significant polluter, their cumulative contribution to levels of environmental degradation may be greater than that of larger companies. For example, it has been estimated that SMEs are responsible for nearly three-quarters of industrial air pollution in Britain (*Financial Times*, 8 January 1996). Compliance with, and enforcement of, environmental regulations may be more lax in smaller companies (Brusco, Bertossi and Cottica, 1996). An effective environmental policy has to apply to SMEs as well as to larger companies, but it should be implemented in a way that does not reinforce the market advantages of larger companies.

WHY WE NEED TO DISTINGUISH BETWEEN SMEs AND LARGE FIRMS

Conceived in terms of a high level of generality, what are the different resource advantages of large firms and small and medium-sized companies in relation to environmental deregulation? It is evident that SMEs operate in a wide range of commercial settings and experience considerable variations in terms of the environmental impact of their operations. Just consider three by no means exhaustive examples: a medium-sized automotive parts company which is supplying to a very limited number of customers; a small courier company with four branch offices which competes with other companies in its locality as well as subcontracting to a larger company; a hairdressing salon which necessarily sells its services on a purely local market. The customer mix of the businesses varies considerably: the hairdressing salon is dealing solely with individual consumers; the automotive parts company is dependent on its contractual relationship with a small number of much larger customers; the courier company survives through a mixture of taking surplus work from a larger competitor, contracting with major local customers and accepting one-off jobs. The automotive parts company may be a significant source of air and water pollution, while the courier company is preoccupied with the health and safety issue of road traffic accidents. If, however, it decides that it wants to be licensed to transport hazardous substances (a value added market segment) it may find itself in a more complex regulatory environment where the 'one line health and safety policy: don't crash' is no longer adequate. The hairdressing salon generates waste materials which may end up in landfills and ultimately leak toxic materials into the aquatic system.

This set of examples is simply a way of demonstrating the point that SMEs considered as individual units of analysis vary considerably, and that often the sectoral distinction will be more important than the large–small firm distinction. This latter point will be returned to more systematically later in the chapter. In order to make some generalisations about SMEs which can assist in modelling the problem, it will be necessary to deal to a large extent in stylised facts. In practice, only some, or perhaps even none, of the problems identified in the following discussion will be experienced by any particular SME. There are also circumstances, which will be discussed, where SMEs may have advantages in the regulatory game over larger firms.

THE DIFFERENTIAL IMPACT OF ENVIRONMENTAL REGULATIONS

Coping with environmental regulations poses special problems for SMEs. One of the most significant problems is the demands that are placed on a managerial team that is limited in size and very hard pressed. A large company, particularly one operating in a sector that is environmentally sensitive, is likely to have a regulatory affairs department with a specialist staff. In an SME, a manager is likely to have to deal with environmental issues alongside a range of other problems. The UNICE Regulatory Report focuses particular attention on this issue:

> Research has shown that one of the major factors determining the success of SMEs is the capacity of management, and that one of the major weaknesses of SMEs is the paucity of management and of managerial skills. Time that is taken up with regulatory issues, cannot be used for directing the strategy and managing the operations of the enterprise. Over the long term, this diversion of scarce managerial resources may prove to be the most damaging of all of the many effects of regulations upon the competitiveness of SMEs.
>
> (UNICE, 1995, p. 29)

The administrative costs of complying with regulations may be dispro-portionately high for SMEs. The UNICE Report quotes evidence from the Netherlands to sustain this point (see Table 9.1). This is not always the case, however, as some regulations may have a differential impact according to the size of firm depending on the nature of the measures taken or the type of activity regulated. For example, the Large Combustion Plants Directive necessarily affects large electricity generating plants. The Drinking Water Directive led large pesticide manufacturers in the UK to fear that 'existing product lines would be placed in some jeopardy and that high standards of enforcement in the UK may drive their activities abroad' (Matthews and

Table 9.1 Administrative costs of regulation in the Netherlands in 1993 (NLG '000)

Number of employees	Charges per company	Per employee
1–9	26	7.6
10–19	44	3.2
20–49	101	3.1
50–99	133	1.9
100 and more	367	1.3

Source: UNICE, 1995

Pickering, 1996, p. 25). In some sectors, there may be no SMEs which are affected by environmental regulations. Even where there are SMEs, they may be able to evade the full effects of the regulations because of inefficient enforcement, so that 'increase of costs will only concern large, highly visible firms, that cannot sneak through the regulations' (Brusco, Bertossi and Cottica, 1996, p. 137).

Where SMEs are present, the fixed costs of installing 'end of pipe' equipment and taking other measures to comply with environmental regulations may bear disproportionately on them. They have less turnover to absorb such capital expenditure and may have less access to financial resources. Admittedly, access to capital for SMEs may vary from one EU Member State to another. Nevertheless, they tend to be 'more reliant upon short-term borrowings than larger companies. SMEs also tend to be unquoted companies ... SMEs tend, therefore, to have less capital available, and hence less capacity to absorb unproductive capital expenditure as a consequence of regulations' (UNICE, 1995, p. 31).

While larger firms are in a better position to exert influence on the drafting of new regulations, SMEs may be in a more advantageous position when it comes to compliance and enforcement. Enforcement is not just a function of the level of activity and effectiveness of enforcement authorities, but also of the vigilance of environmental activists and citizens who draw attention to failures to comply with regulations. It may be that 'larger, more visible firms are more closely watched by the public than small, local businesses' (Brusco, Bertossi and Cottica, 1996, p. 124). In their analysis of the European waste disposal industry, Brusco, Bertossi and Cottica suggest that small local firms and taxpayers may collude to tolerate lax enforcement. Small local firms may be able to exert political influence on the process of enforcement. Brusco, Bertossi and Cottica conclude (1996, p. 140) 'that small firms have captured not regulation, but enforcement and control, and in this way they [have] contrasted the growing power of larger players'.

Apart from the tendency of the enforcement authorities to devote the greater part of their resources to larger firms, these firms are also more likely to face pressure from stakeholders such as customers, shareholders and residents of the area near the plant to improve environmental performance. Research by the Association of British Chambers of Commerce suggests that only one in five of smaller companies in Britain have encountered pressure from their customers to improve environmental performance. A survey of five hundred small businesses by the North West Regional Association Environmental Group found that 97 per cent of respondents had not conducted any form of environmental audit; a similar percentage had no environmental policies, while 75 per cent offered no environmental training (*Financial Times*, 8 January 1996). Similar surveys in other Member States might produce rather different results, but the underlying point remains that smaller companies face a rather different incentive structure in terms of responding to demands from their customers and shareholders on environmental issues.

Of course, even in larger companies, regulatory affairs divisions may be concerned primarily with 'damage avoidance' and 'compliance cost reduction' rather than with developing a positive corporate environmental policy. Such a policy may be counter-productive as it overlooks the extent to which regulation may serve as a stimulus to innovation. The approach within companies, however, is often to regard regulations as a nuisance to be managed through a minimal compliance with legal requirements. Research by Fineman on British companies found that the managers interviewed who had special responsibility for environmental affairs differed little from their colleagues in terms of their relatively limited commitment to environmental protection. In some firms, environmental affairs were an extension of the responsibility of middle level technical managers for health and safety issues. The functional segmentation and cost centredness of many companies meant that many suggested environmental improvements met internal resistance and had to be negotiated in internal decision-making bodies. 'All companies wished to avoid the public embarrassment of pressure group "revelations". These, together with the fear of being prosecuted, were a potent influence on an organisation's environmental actions' (Fineman, 1995, p. 6). Concerns about corporate image are, nevertheless, better developed in larger companies than in SMEs and Fineman did find larger companies with strong corporate cultures which embodied a commitment to environmental initiatives.

For the larger company in an oligopolistic sector, environmental regulations may be welcomed as a form of entry barrier, particularly if they can also serve as an effective non-tariff trade barrier. If large firms wish to use environmental regulations as an entry barrier, they are well placed to do so. As Hancher and Moran (1989, p. 272) note in the conclusion to an edited collection on regulation, 'The importance of the large firm in the regulatory process is particularly notable. Indeed, an important theme of the contributions to this volume is the central place of the large, often

151

multinationally organised, enterprise as a locus of power, a reservoir of expertise, a bearer of economic change, and an agent of enforcement in the implementation process.'

What the incentives are for large firms to use environmental regulations as an entry barrier is open for debate. It seems probable that they are most likely to be used in this way when there are issues of quality control affecting competitiveness. Larger firms may see themselves facing a free-rider problem in the form of price competition from 'cowboy firms' which are neglectful of the environmental impacts of their activities. As well as offering 'unfair' price competition such firms may damage the image of the industry with the customers and wider public. However, the real problem here may be one of enforcement as 'cowboy' firms may disregard or flout environmental laws, sometimes in complicity with large companies as in the case of the illegal dumping of toxic wastes.

In many cases it may not be necessary to erect entry barriers because small firms do not present a direct challenge to larger firms. Either they do not exist at all in the sector (automotive industry) or they are serving specialist niche markets (large scale dairy processors compared with farmhouse cheesemakers). In other cases, the smaller firms may be significantly dependent on the larger firms because of the nature of their contractual relationship (automotive parts). Indeed, supply chain management may be becoming a more central feature of the organisation of modern manufacturing processes, creating new relationships of mutual dependence between large firms and SMEs:

> Enhanced competition for markets for end products and more insistent demands for high quality, fast and timely delivery and more diversified products have called for significant changes in the ways in which industrial firms organise their production process. The above demands can no longer be fulfilled by isolated organisations but only through the development of close cooperation between manufacturing firms and their suppliers.
>
> (Lane and Bachmann, 1995, p. 1)

In certain instances, it is possible that larger firms might insist on environmental standards being met in their supplier firms alongside other requirements such as meeting 'just in time' production requirements (the British retail firm, Marks & Spencer, has for a long time insisted on certain welfare standards being met by its suppliers). In this kind of supply chain relationship, the smaller firm is, in effect, 'captured' by the larger firm and is subordinated to its requirements. Indeed, this possibility is given some emphasis in the EU's action programme on the environment: 'many [SMEs] survive on the demand created by large firms which will be obliged to tighten up their processes and meet the overall criteria of the ecological labelling system' (European Community, 1993, p. 64).

Smaller firms are not just suppliers to larger firms, but also their customers. Some large firms may have an interest in strengthening environmental regulation because it enlarges the market for their products. Smaller firms may have to purchase 'end of pipe' technology from larger firms. In Britain at least, however, the market for environmental technology seems to be characterised by small suppliers and big purchasers. For example, Degremont UK which supplies water and effluent treatment products with 80 per cent of its customers being the privatised water companies and the remaining business coming from multinationals such as BP, ICI and Unilever. Envirosystems, which produces a specialist gas monitor, has Zeneca and Rolls-Royce among its principal customers. The Environment Industries Commission, the UK trade association for the sector, sees the main demand for environmental control equipment coming from the largest five hundred UK companies by market capitalisation (*Financial Times*, 21 June 1995).

The market is certainly a substantial one. The first UK survey on the subject found that 87 per cent of the companies contacted estimated that their expenditure on such technology would increase before the end of the 1990s. Three-quarters of the respondent companies expected to spend up to 50 per cent more on such equipment before the year 2000, although they had already increased expenditure on environmental technology in the first five years of the decade (*Financial Times*, 21 June 1995). At an international level, the OECD predicts that the demand for waste management products and air pollution control equipment will grow by over 50 per cent in the 1990s with water and effluent treatment products growing by a third. Other analysts such as the US Environment Protection Agency consider that these predictions underestimate likely demand.

THE ECONOMIC AND POLITICAL VULNERABILITY OF SMALLER FIRMS

Although the growth of environmental technology markets may create new opportunities for SMEs, the general point that needs to be stressed is that smaller firms are economically and politically more vulnerable than larger firms. They have smaller financial reserves, and less access to financial support in times of difficulty than larger firms. Their management teams are much smaller, have less specialist expertise, particularly on environmental questions, and are more stretched than in larger firms. They may be dependent on large firm customers for much of their business and may be enmeshed in increasingly demanding supply chain relationships. If they are directly competitive with large firms, they may be driven out of business by predatory pricing. If all else fails, larger businesses may simply buy up their small competitors. In sectors where family ownership may have customer care advantages, large chains have made use of their financial power to acquire businesses and to seek economies of scale, e.g. the funeral business in both Britain and the United States.

Smaller firms may be able to find market niches in which they are able to survive and even flourish, either as suppliers to larger companies or serving customers directly. However, if the competitive arena is a difficult one for SMEs, it is even harder for them to penetrate the policy arena. In political terms, large firms are in a much stronger position. As Lévêque notes (1996b, p. 21), 'The pre-eminence of large firms in the regulation devising process is observed at all levels.' Very large firms are able to afford their own government relations or European affairs divisions in addition to regulatory affairs offices. Coen's research (1996) shows that environmental policy is the most Europeanised of all the issue domains of interest to firms so a capacity to exert effective influence at the European level is particularly important. Large firms are able to afford to give substantial donations to political candidates or parties in those countries where such donations are viewed as acceptable. They may seek to shape the nature of the debate about environmental questions by 'colluding to finance scientific interpretations that are contrary to established [Intergovernmental Panel on Climate Change] viewpoints' (O'Riordan and Jordan, 1996, p. 79).

Within trade associations, the interests of large firms tend to predominate. They pay the subscriptions which allow the association to function effectively and lend their senior and technical staff to provide its committees with essential expertise which is fed into the regulatory decision-making process. Thus, the environmental affairs committee of a trade association may be made up largely of the environmental specialists of the principal companies in the industry. In the case of the European Chemical Industry Council (CEFIC), direct membership of the federation is limited to forty-nine leading companies, with SMEs being represented through their national federations. Even when they are full members of European associations, SMEs may find it difficult to have an impact on the decision-making process within the organisation:

> in the waste management regulatory process, trade associations have always spoken out in representation of medium-sized and, above all, large firms. This is a feature shared with most industries; even the Italian craftsmen's associations, which account for more than half the market in the industries they work in, find it difficult to make their voice heard in Brussels.
>
> (Brusco, Bertossi and Cottica, 1996, pp. 137–138)

In the waste management sector, the smallest, locally based firms 'are not represented in the European regulatory arena nor, generally speaking, in the national ones' (Brusco, Bertossi and Cottica, 1996, p. 138). In some cases, entry barriers may exclude smaller firms. The UK Producer Working Group on compliance with the packaging Directive was set up at government instigation to represent a wide range of affected interests, but participation was dependent on the payment of a very large subscription. In some industries there are specialist organisations for SMEs, but they tend to be less influential and less well resourced than the organisations representing larger companies. The lack of understanding which may exist between larger and smaller companies within a trade association is illustrated by the example of the EU's Eco-Management and Audit Scheme (EMAS). Smaller companies are less likely to participate in this voluntary scheme (Franke and Wätzold, 1996). The German Verband der Chemischen Industrie (VCI) has pointed out that a number of German chemical companies have successfully piloted the scheme. It 'expects the vast majority of its members to pursue the audit as it provides a method of cataloguing information and a good view of a company's own organisation' (*European Chemicals News*, 17–23 April 1995, p. 30). However, the example the VCI gives is of a large company plant, the Bayer site at Dormagen. It is open to question how effective organisations like the VCI are in articulating the views of their smaller company members. Comments such as those by the environmental vice president of a leading chemical firm that smaller companies who cannot afford to pay consultants to conduct audits should try to learn to do the audits themselves (*European Chemicals News*, 20 June 1994, p. 25) betrays a lack of understanding of the operating environment faced by SMEs.

Given all these disadvantages, one might ask how smaller businesses survive? A number of strategies are possible. Some personal services have to be provided at the point of sale and there is little advantage in larger scale organisation, e.g. hairdressing. In some cases, the total market may be too small to interest larger firms. In many instances, SMEs are able to compete effectively because they have lower overheads and are therefore highly price competitive. Perhaps one of the greatest advantages of the SMEs is their flexibility, exemplified by the debate on 'flexible specialisation' initiated by Piore and Sabel (1984). 'Flexible specialisation' was presented as 'a new phase of capitalist production characterised by craft labour, small-scale industry using the latest technology, and diversified world markets and consumer tastes' (Murray, 1987, p. 84). According to UNICE, 'The European regulatory framework with its enormous scope and prescriptive nature, reduces the flexibility of all companies. This has a disproportionate effect on to SMEs' (UNICE, 1995, p. 30).

The general conclusion to be drawn is that larger firms are better able to cope with the negative effects of regulation. The market for environmental technology may, however, open up new opportunities for smaller firms. Environmental regulation may also curb the wasteful use of inputs such as energy in the production process and lead to a search for more effective ways of utilising by-products. Larger firms are more likely to have the research and development capacity to benefit from such opportunities.

155

Against this, it should be noted that the director of the UK Environment Industries Commission considers that 'Most businessmen still see the environment as a cost rather than an opportunity to improve efficiency' (*Financial Times*, 21 June 1995).

SECTORAL DIFFERENCES

These generalisations have to be qualified by the existence of significant differences between sectors in terms of the environmental impact of their activities. Some industries, notably chemicals, cement, tanning, power generation and pulp and paper production, have a substantial and immediate impact on the environment. These directly polluting industries often have relatively mature technologies. Others have a significant indirect effect on the environment, notably the automotive industry. Yet other industries rely on the use of substantial amounts of packaging which has increasingly become the subject of regulatory measures, e.g. the 'fast moving consumer goods' industries such as food processing, cosmetics and toiletries. Other industries may, however, have a more limited environmental impact (e.g. many service industries) or may be seen as having environmental impacts which are beneficial in net terms (e.g. the manufacture of light rail equipment, environmental control technology).

It may be, therefore, that both large and smaller chemical companies have more in common in terms of environmental regulation than, say, a small chemical company and a small clothing company. Sectoral differences may override those based on size. However, one should not push this point too far. If one compares the results on the impact of regulation on competitiveness for the UNICE regulatory survey as a whole with those for the chemical industry (CEFIC, 1996), some differences emerge, but they are not very substantial. More chemical companies did designate environmental regulations as 'very unhelpful' or 'unhelpful', but only by a few percentage points. Chemical companies were also more likely to find regulations or technical standards 'helpful' compared to companies in general, but again only by a few percentage points. 'There is a high degree of consistency as to which areas of regulation have the greatest adverse impact on competitiveness. SMEs and large companies, as well as companies in manufacturing and services, and companies in different countries, all identify [tax and other administrative law, employment law and the environment] as having the greatest negative impact' (UNICE, 1995, p. 32).

The UNICE study identified four EU regulations perceived as having the greatest adverse effect on their competitiveness: the Framework Directive on Waste (91/156); the Packaging and Packaging Waste Directive (94/62); the Discharge of Dangerous Substances into Water Directive (76/464); and the Environmental Impact Assessment Directive (85/337). All these are 'generic' Directives which apply across all sectors of the economy unlike, for

156

example, the 'Seveso' Directive on Major Accident Hazards which had a particular impact on the chemical industry. The EU Environmental Action programme distinguishes between sectors in fairly broad brush terms: manufacturing industry; energy; transport; agriculture; and tourism (European Community, 1993).

In order to understand the dynamic behind moves towards deregulation, one has to understand how regulation impinges on companies. The discussion so far has thus been related to regulation rather than deregulation. The central question has been to what extent have large firms and SMEs been differentially affected by the growth of regulation. The general answer has been that larger firms are better placed to cope with environmental regulation which may confer on them some competitive advantages. However, in spite of having invested considerable money and effort to develop effective internal systems of compliance with environmental regulations, and also having in some cases defined their corporate image in terms of being 'environmentally friendly', this does not mean that large firms necessarily oppose deregulation. In order to understand the move towards environmental deregulation, it is first necessary to review developments in this direction in the United States. It is evident that these have had a direct impact on the debate in Europe. For example, 'Taking its cue from the Chemical Manufacturers Association in the US, the German chemical association, VCI, thinks things are now moving in the same direction in Germany' (*European Chemical News*, 17–23 July 1995, p. 21).

DEREGULATION IN THE UNITED STATES

The United States has played an important agenda setting role in environmental policies. Reflecting on the findings of their comparative analysis of regulation in Europe and the United States, Brickman, Jasanoff and Ilgen conclude (1985, p. 300):

> the United States undeniably plays a leadership role in the cross-national alignment and evolution of chemical control policies. [Particular] features of the American regulatory setting – the openness of decision making, the heavy investments in regulatory programs and on health effects research, and a political process that feeds on controversy – serve to uncover and dramatise the hazards of chemical technology. They also produce policy initiatives . . . which tend to arise more slowly, if at all, in European regulatory settings.

Within the United States, California has played a significant agenda setting role, pioneering a number of innovations in environmental policy. However, in the early 1990s California was afflicted by a global recession which was deepened in the state by the end of the Cold War which led to a major rundown of the defence industry. As the recession deepened, voters

157

apparently became less sensitive to environmental concerns. For example, in 1990 they approved financing measures for rail transport projects; in 1992, the second phase was defeated by a narrow margin; and they went down heavily on a third vote in 1994. The defeat of Kathleen Brown in the 1994 election for governor, and the eventual seizure of control of the Assembly by the Republicans in 1996, all pointed to a shift in the state's political balance. This shift in political sympathies had significant consequences for environmental policy. Acting under pressure from the Governor's office, the predominantly Republican Air Resources Board concluded a deal with the auto industry in February 1996 which effectively suspended the state's innovative electric vehicles programme.

At a federal level, the chemical industry has been in the forefront of pressure for what has been termed by them a most responsible cost–benefit risk-based approach to regulation. The drawback with such an approach is that environmental gains are often less readily quantifiable in terms of cost–benefit analysis than the gains accruing to industry. The then chairman of the Chemical Manufacturers' Association claimed in May 1995 that every one dollar increase in the cost of pollution abatement reduced productivity by between three to four dollars. Thus:

> Regulatory reform has surfaced with a vengeance in the US. The issue is close to the heart of chemical industry executives, who have long argued that a climate of over-regulation has placed a drag on their industry. But the recent lean times, which forced . . . downsizing and restructuring, also served to drive home a recognition that resources are not infinite and that the US 'command-and-control' regulatory system may not make the best use of those resources.
>
> (*European Chemicals News*, 7–13 August 1995, p. 22)

These concerns have been picked up by President Clinton in his 'Reinventing Government' initiatives. The broad objectives of this programme have been defined by the President in the following, possibly somewhat contradictory, terms:

> I want a government that is limited but effective, that is lean but not mean, that does what it should do better and simply stops doing things that it shouldn't be doing in the first place, that protects consumers and workers, the environment, without burdening business, choking innovation or wasting the money of the American taxpayers.
>
> (The White House, 1995)

In March 1995 President Clinton and Vice-President Gore unveiled the 'Reinventing Environmental Regulation' initiative designed to encourage collaboration between industry and regulators. The President suggested that more use should be made of market mechanisms, extending the entitlement of utilities to buy and sell rights under the Clean Air Act to other areas of clean air and water protection. Compliance centres would be opened to help small businesses who would have six months to solve problems without being fined. The President praised a pilot EPA project called Project XL which he described as 'here is the pollution reduction goal. If you can figure out how to meet it, you can throw out the EPA rulebook' (The White House, 1995). There has been a similar shift away from a command-and-control approach in Europe, although 'EU environmental regulatory reform is a recent, on-going process' (Lévêque, 1996b, p. 18).

The EPA, which was already working on its own 'Common Sense Initiative', has sought to implement these projects. The director of EPA's regulatory reinvention scheme has emphasised that the EPA is 'trying to introduce flexibility with stakeholder involvement'. But he also stressed, 'None of the reinvention projects are rolling back our commitment to environmental quality or standards' (*European Chemicals News*, 7–13 August 1995, p. 22). Other work is being undertaken by the President's Council on Sustainable Development which 'I established to outline a new direction in environmental policy' (Statement from Office of the White House Press Secretary, 7 March 1996). It is seen by the President as a consensus building mechanism between business, environmentalists and government in an era of smaller government.

In late 1995 and early 1996 the question of environmental deregulation got caught up in the budget battle and the election year atmosphere. The budget that the House passed for the EPA in August contained a 34 per cent cut in EPA resources, and a 50 per cent cut in enforcement dollars. The EPA appropriations bill also included eighteen riders affecting the EPA's ability to act and to enforce environmental laws and regulations. Taken together, these prohibitions would 'essentially shut down the Clean Water Act' (Briefing by EPA Administrator Carol Browner, White House Briefing Room, 8 August 1995). The bill President Clinton vetoed in December cut EPA funding by 22 per cent. By March the Senate was proposing to restore a fraction of this funding and the House none at all. The legislation still contained what Vice-President Gore described as 'special interest riders and deep budget cuts that would roll back protections for America's air, water and public lands . . . They sent us these same extreme measures on the environment . . . Yes, the era of big government is over . . . But let's do it in a way that . . . protects our environment' (Office of the Vice-President, 13 March 1996).

What became evident was that there was still considerable support among the American population for the maintenance of standards of environmental protection, and the subject became a recurrent theme in President Clinton's campaign speeches. The passage of the Safe Drinking Water Act in August 1996, admittedly too late to release the appropriate funds, was hailed by the President as 'a model for responsible reinvention of regulations. It replaces an inflexible approach with the authority to act on contaminants of greatest risk and to analyse costs and benefits, while retaining public health as the paramount value' (Statement by the President, Office of the Press Secretary, 6 August 1996).

THE DEBATE IN EUROPE

The new approach being developed in the United States had an impact on the debate in Europe, although the pressure for deregulation also had European origins. For example, in 1993 the influential European Round Table of Industrialists, made up of the chief executives of leading European companies, called for significant changes to the extent and nature of the European regulatory framework as part of their programme to restore European competitiveness.

This new thinking permeated the Commission. In a speech in 1994, Bernard Delogu, a leading DG XI official in charge of environmental auditing and control of industrial installations, stated that future European environmental legislation would be strongly influenced by the concept of market forces and less dependent on 'command-and-control' regulation. He cited the EU's voluntary schemes for eco-labelling and environmental management and auditing (EMAS) as examples of this new trend (*European Chemicals News*, 20 June 1994, p. 25). Some concern was expressed by CEFIC's technical counsellor that if EMAS became a requirement for companies, its toll on costs and manhours could push smaller ones out of business, although, as designed, the scheme was flexible enough to enable smaller firms to cope.

What particularly drove the debate forward in the EU was concern about the erosion of German competitiveness (for example, the report of the 'Five Wise Men' in November 1994 argued that any export-led recovery must be coupled with more deregulation). In April 1994, the British and German governments set up a group of business leaders as part of an effort to curb the effects of what was described as over regulation. Its UK chairman was a former government minister, Francis Maude, who was a member of the deregulation task force sponsored by Michael Heseltine. The German membership included Jürgen Strube, the chairman of BASF. In the spring of 1995 this group called on the Commission to scrap the Directive on Dangerous Substances, revise the Drinking Water Directive and amend proposals for Integrated Pollution Prevention and Control.

This alliance between Britain and Germany to press the case for deregulation surfaced again at the Corfu European summit in June 1994 when Britain supported Germany's request for the establishment of a group of businessmen and civil servants to examine whether EU and national legislation was imposing unnecessary burdens on companies. This led to an establishment of a group of seventeen 'wise men' known as the Molitor Committee, chaired by Bernhard Molitor, a former senior official in the

German economics ministry. The committee's terms of reference were to examine national and EU-wide legislation and their effects on competitiveness and job creation and to recommend how regulations could be abolished or simplified.

Its report was presented to the Cannes summit in June 1995. The report did not examine national regulations because of a lack of time and only considered EU regulations. It did not call for widespread deregulation, but stated that regulatory frameworks must be reviewed if competitiveness and employment goals were to be achieved. It did identify environmental and other regulations which were creating additional burdens for companies.

However, the report managed to offend most of the key constituencies to which it had tried to appeal. Jacques Santer pointed out that most of the burdens on industry stemmed from national legislation. The Commission defended the benefits of regulation, arguing that 'A good regulatory framework sets out the field within which the businessman knows he can operate freely' (*Financial Times*, 23 June 1995). The European Trade Union Confederation attacked the report's conclusions on the grounds that they undermined social policy. Describing the report as 'flawed with few friends', the ETUC stated that it was 'deeply perturbed by the group's findings' (*Financial Times*, 19 June 1995). Nor were the employers satisfied. Sir Michael Angus, the chairman of Whitbread, backed by employers' representatives from Germany, Ireland and the Netherlands produced a minority report critical of the social policy recommendations.

CONCLUSIONS

What the consequences are of the pressure for environmental deregulation in both Europe and the United States remains to be seen. Budgetary constraints may ultimately have a greater influence on environmental regulation in the United States than in Europe. One suspects that the objectives of the participants in the policy process are still somewhat different despite all the talk of seeking consensus. Environmental agencies would no doubt like to have a better relationship with business, and would probably be prepared to settle for an approach to environmental regulation which focused on outcomes rather than the means used to secure those outcomes. However, this does not mean that environmental agencies share the business goal of 'a better normative "culture" at EU, national, regional and local levels to encourage lawmaking based on sound cost benefit analysis' (CEFIC, 1994, p. 15). What that would mean in practice would be that some environmental regulations would be disallowed because the net benefits (environmental gains versus economic costs) would be too small.

The 1993 EU environmental action programme considers the special problems of SMEs, but is relatively unsympathetic to them. The report argues that there should not be discrimination in favour of SMEs because small plants cause their share of pollution and waste. Some variations in the time-frames for implementation might be possible to help SMEs, and they could be provided with expert services and training programmes. However, it is argued that their size allows them to be more adaptable, while adherence to state of the art requirements could give them a market edge (European Community, 1993, p. 64).

The UNICE Regulatory Report, published after that of the Molitor Committee, places particular emphasis on the regulatory hurdles faced by SMEs. Eighty per cent of the 2,100 companies surveyed for the report were SMEs. Examining the effect of regulations on research and development in 'building block' technologies such as biotechnology and new chemical compounds, the report argued that regulations led to the use of existing technologies rather than the exploitation of new ones. It was argued that 'the long-run effects will be to cede to firms outside Europe "first mover" advantages and to favour large, multinational companies at the expense of SMEs'. Multinational companies outside Europe would be able to introduce new biotechnology products into the European market as soon as European regulations changed. 'European-based SMEs will not have had experience of such products and will be unable to respond quickly, leading to losses of market position' (UNICE, 1995, p. 20). The report argued that the 'first order' costs of regulations for SMEs were serious enough in terms of increased administrative costs and the use of management time. However, 'the second order effects may be even more serious: fewer new products, fewer new SMEs, and the slower growth and increased failure of existing SMEs' (UNICE, 1995, p. 45).

The proposals made in the UNICE Report to help SMEs, such as a programme to reduce compliance costs, are well intentioned but one has to ask whether they are likely to be put into effect. Business is a very influential actor within the EU, but the balance of power lies with big business. The single market as a concept is designed for large multinational companies. Cowles summarises (1995, p. 523n) the special advantages of multinationals when they mobilise politically at a supranational level:

> First, unlike domestic interests that may not be aware of EU legislative action, MNEs organised at the EU level have the means and information to set or influence the policy agenda and/or shape the policy alternatives long before government leaders come to the negotiating table. Second, these same firms can influence the votes of government leaders by meeting directly with high-level government officials in Brussels and in the nation-states. Third, the companies can help shape the preference formation of groups located at the domestic level.

SMEs and larger firms face very different operating environments. Being a manager in an SME is very different from being a manager in a large multinational company. SMEs are more concerned with reducing the general volume of regulation. Larger companies are more concerned with the form of regulation and may see certain types of regulation as giving them competitive advantages. They have a range of options in terms of regulating their relationships with SMEs, with supply chain management becoming increasingly important. As well as having less economic power, SMEs also have less political influence. Environmental deregulation may therefore occur in a way which benefits large firms. Depending on how it was implemented it could even reduce the market for environmental control technology which is an area in which SMEs have been active. One could end up with a suboptimal outcome in which SMEs were less competitive and environmental protection less secure.

ACKNOWLEDGEMENTS

The author would like to thank Duncan Matthews, Ute Collier and Jonathan Golub for commenting on earlier versions of this chapter (they bear no responsibility for its final contents) and Ros Grant for sharing her experience of being a manager in a small business.

REFERENCES

Brickman, R., Jasanoff, S. and Ilgen, T. (1985) *Controlling Chemicals: the Politics of Regulation in Europe and the United States*, Ithaca: Cornell University Press.

Brusco, S., Bertossi, P. and Cottica, A. (1996) 'Playing on two chessboards – the European waste management industry: strategic behaviour in the market and in the policy debate', in Lévêque, F. (ed.) *Environmental Policy in Europe*, Cheltenham: Edward Elgar.

CEFIC (1994) *Annual Report*, Brussels: European Chemical Industry Council.

CEFIC (1996) 'The UNICE regulatory study: findings from the chemical industry', typescript presentation, Brussels, Council of European Chemical Industry.

Coen, D. (1996) 'The large firm as a political actor in the European Union', unpublished Ph.D. thesis, Florence: European University Institute.

Cowles, M.G. (1992) 'Setting the agenda for a new Europe: the ERT and EC 1992', *Journal of Common Market Studies*, 33 (4), pp. 501–526.

European Community (1993) *Towards Sustainability*, Luxembourg: Office for Official Publications of the European Community.

Fineman, S. (1995) 'The greening of management: a study of moral belief and action', unpublished report to Economic and Social Research Council.

Franke, J.F. and Wätzold, F. (1996) 'Voluntary initiatives and public intervention – the regulation of eco-auditing', in Lévêque F. (ed.) *Environmental Policy in Europe*, Cheltenham: Edward Elgar.

Hancher, L. and Moran, M. (1989) 'Organizing regulatory space', in Hancher, L. and Moran, M. (eds) *Capitalism, Culture and Economic Regulation*, Oxford: Clarendon Press.

Ikwue, A. and Skea, J. (1996) 'The energy sector reponse to European combustion emission regulations', in Lévêque, F. (ed.) *Environmental Policy in Europe*, Cheltenham: Edward Elgar.

Lane, C. and Bachmann, R. (1995) *Cooperation in Vertical Inter-Firm Relations in Britain and Germany: the Role of Social Institutions*, Working Paper 21, ESRC Centre for Business Research, Cambridge: University of Cambridge.

Lévêque, F. (1996a) 'Conclusion', in Lévêque, F. (ed.) *Environmental Policy in Europe*, Cheltenham: Edward Elgar.

Lévêque, F. (1996b) 'The European fabric of environmental regulations', in Lévêque, F. (ed.) *Environmental Policy in Europe*, Cheltenham: Edward Elgar.

Matthews, D. and Pickering, J. (1996) *The Role of the Firm in the Evolution of European Environmental Rules: the Case of the Water Industry and the European Drinking Water Directive*, Discussion Paper 92, London: National Institute of Economic and Social Research.

Murray, F. (1987) 'Flexible specialisation in the "Third Italy"', *Capital and Class*, 33, pp. 84–95.

O'Riordan, T. and Jordan, A. (1996) 'Social institutions and climate change', in O'Riordan, T. and Jäger, J. (eds) *Politics of Climate Change: a European Perspective*, London: Routledge.

Piore, M.J. and Sabel, C.F. (1984) *The Second Industrial Divide*, New York: Basic Books.

UNICE (1995) *Realising Europe's Potential through Targeted Regulatory Reform: the UNICE Regulatory Report*, Brussels: UNICE.

The White House (1995) 'Remarks by the President at Rego Event, Custom Print, Arlington, Virginia', Office of the Press Secretary, 16 March.

164

10

ENVIRONMENTAL POLICY INSTRUMENTS IN A DEREGULATORY CLIMATE

The business perspective

Giorgio Porta[1]

INTRODUCTION

There is a widespread understanding today between governments and companies that environmental protection must be among the highest priorities of every business (International Chamber of Commerce, 1991). Whereas until the early 1990s, the environment was a concern mainly for European and national legislators, it is now part of the general preoccupations of business. The question is however which instruments are most suitable for achieving a high level of environmental protection.

Two reports were recently published dealing with the impact of (environmental) regulations on competitiveness: the UNICE Regulatory Report, elaborated by the environment working group of the Union of Industrial and Employers' Confederation of Europe (UNICE, 1995), and the Molitor Report (European Commission, 1995a). This chapter focuses on the findings of these two reports. The reports show on the one hand that companies accept adequate governmental intervention in the environmental field, but on the other hand that there are many defects in regulation, which lead to increased operating costs, capital expenditure and diversion of management time.

Governments and companies accept a shift in responsibilities for environmental policy. Companies take their own responsibilities by setting up and implementing environmental management systems, voluntary approaches and negotiated agreements. In this framework, they accept as well that legislation is no longer the only, but one of many instruments for achieving a high level of environmental protection. Many examples show that this process of working together is developing, although at a different pace from one country to another.

The need for Sustainable Industrial Development (SID) is increasingly recognised by business. To achieve eco-efficiency, new instruments are needed. One of the important new instruments to develop and achieve SID is the negotiated agreement. The European Commission is currently working on a Communication concerning negotiated agreements. In this chapter attention is paid to the nature of such agreements, their potential benefits, the conditions to make agreements work and their possible contents.

EMPHASIS ON LEGISLATION

As a rule, environmental policy within the EU is implemented by legislation. The Single European Act of 1 July 1987 amended the Treaty of Rome to make two explicit references to the Community's powers in the field of environmental protection. Article 100a seeks to harmonise conditions of trade and competition within the European single market. It is employed primarily for setting uniform standards for traded products. Indirectly Directives and regulations based on Article 100a can serve environmental protection. Examples are standards for exhaust emissions from vehicles, restrictions on marketing and use of dangerous chemicals, and requirements for packaging and packaging waste.

Article 130s forms part of the environmental policy chapter of the Treaty. It is the legal base for legislation the sole aim of which is environmental protection. Examples of legislation on this basis are nature protection, environmental management (EMAS), water quality and air quality management.

In 1992, the Treaty of Maastricht, founding the European Union, introduced as a principal objective the promotion of sustainable growth respecting the environment (Article 2). It includes among the activities of the Union a policy in the sphere of the environment (Article 3 (k)), specifies that this policy must aim at a high level of protection and that environmental protection requirements must be integrated into the definition and implementation of other Community policies (Article 130R.2).

Over the past two decades four EU action programmes on the environment have resulted in hundreds of Directives and regulations covering pollution of the atmosphere, water and soil, waste management, safeguards in relation to chemicals and biotechnology, product standards, environmental impact assessments and protection of nature.

Previous environmental measures have tended to be prescriptive in character, with an emphasis on the 'thou shall not' rather than the 'let's work together' approach. As a consequence, there has been a tendency to view industrialisation or economic development and environmental concern as being mutually hostile, and not as an incentive for shared responsibility in the environmental and/or the economic field (European Commission, 1992).

THE IMPACT OF REGULATION ON COMPETITIVENESS

Companies need adequate environmental legislation to establish a framework which facilitates business activity and confidence. However, at the same time legislation at Community and national level can inhibit companies' capabilities to create employment and improve business dynamism. This adverse impact can result from the cost and uncertainties created by legislative complexity and rigidity, disproportionate administrative burdens and impediments to innovation. Over-regulation stifles growth, reduces competitiveness and costs Europe jobs (European Commission, 1995a).

This analysis of the impact of legislation on competitiveness was developed in two reports published in 1995: the UNICE Regulatory Report and the Molitor Report. The findings in the UNICE Report are based on a wide-ranging examination of relevant literature, workshops and a survey of 2,100 companies (80 per cent of which are SMEs and 40 per cent in the service sector) in 14 countries which examined the effects of regulations on competitiveness.

The UNICE Regulatory Report shows that companies consider that government regulations constrain their ability to improve their competitiveness. It is likely that regulation will become a greater constraint for European companies in the future. Changes in the global business environment will require faster innovation and more flexible operating methods. In addition, the relative position of European companies will decline as other OECD countries lighten their regulatory burdens.

The Molitor Report commissioned by the European Commission from an independent group of experts gave the warning that Europe cannot ignore the fact that other industrial countries with which it competes are making strenuous efforts to reduce their own regulatory burdens.

Companies reported to UNICE that the areas of regulation which have the greatest adverse effect on their competitiveness are tax and other administrative regulations (77 per cent of companies), employment regulations (70 per cent) and environmental regulations (57 per cent). Companies say that in general there are too many regulations (79 per cent), they are too complex (55 per cent), they change too frequently, they are enforced inconsistently and standards laid out in the regulations differ between countries.

To understand the detailed problems that companies experience with regulations, UNICE examined, with company experts, a number of specific European regulations in the areas of the Environment and Health and Safety. The five most common 'defects' across nine environmental regulations are:

- regulations are not proportionate to hazard and risk;
- regulations are too complex or prescriptive;
- enforcement is inconsistent between countries;
- technical definitions are poor or inconsistent;
- regulations are too difficult to understand.

Companies identified four regulations as having, or likely to have, the greatest adverse effect on their competitiveness. These were: the Framework Directive on Waste (91/156), the Packaging and Packaging Waste Directive (94/62), the Discharges of Dangerous Substances into Water Directive (76/464) and the Environmental Impact Assessment Directive (85/337). The main impacts upon firms are that regulations lead to increased operating costs, increased capital expenditure, and diversion of management time.

According to the UNICE Report, many aspects of the current regulatory regime in Europe protect the 'status quo', and reduce the competitiveness of European companies in world markets. Its extensive scope and rigid nature tends to discourage new entrants, to impede the use of new methods of production and to inhibit the exit of existing competitors. It also increases costs, reduces operational flexibility, distorts capital expenditure and diverts the use of management time. Hence it dilutes the level of competitive intensity and impedes the rate of growth in companies, particularly SMEs.

Regarding the principal effects of the regulatory framework on the competitiveness of European companies, the UNICE Report matches the conclusions of the Molitor Report. These principal effects are:

- competitive intensity is diluted;
- barriers to innovation are created;
- obstacles to improving operating efficiency result;
- the development and growth of SMEs is impeded;
- structural adjustments are delayed.

POSSIBLE IMPROVEMENTS

The UNICE Report shows that companies are in favour of a range of improvements to environmental regulations. In many cases they are seeking less regulation (*deregulation*). However, they express an even stronger wish for targeted changes to improve the quality and harmonisation of regulations (*simplification*) and their enforcement.

Overall, they do not see one solution. Companies appear to recognise not only the complexity of the problem but also the need for stability in the regulatory framework. Companies ask for a framework which is supportive of competitiveness, promotes consumer confidence and facilitates the pursuit of other social and economic objectives.

Overall, the UNICE Report calls for the adoption of a three-stage action programme to achieve this new framework:

- a targeted reform of the existing framework;
- a change in the process by which governments intervene in the activities of companies;

168

- a change in the attitudes of society and government to the use of regulation.

The UNICE Report and the Molitor Report mention more or less the same priorities for reforming existing and future environmental regulations. The UNICE Report suggests that government intervention throughout the EU should be goal-based, with specific goals and objectives being consistent across the EU and other OECD countries (particularly the US and Japan). There should be a comprehensive review of all the regulations concerning 'waste', taking account of the findings of the UNICE surveys.

The Molitor Report advocates a new approach to environmental regulation, which stresses the setting of general environmental targets[2] whilst leaving the Member States and, in particular, industry the flexibility to choose the means of implementation; this should be pursued vigorously, and should be the basis for a full-scale phased review of existing environmental legislation (proposal 1). The implementation of policies aimed at broad environmental goals should, where appropriate, approach the environment through the integrated chain management of substances, focusing on inputs, process, waste, emissions, and the consumption and disposal of the final output (proposal 4).

The UNICE Report suggests that regulations should be developed using sound scientific analysis (including risk assessment), and cost–benefit analysis, within a consultation process which is more transparent. According to the Molitor Report, policy should, wherever possible, be designed to achieve a required level of environmental quality, bearing in mind available technology; balancing known emissions with the carrying capacity of the environment, and minimising leaks such as uncontrolled waste or fugitive emissions (proposal 2).

Furthermore, proposals should not be brought forward unless the cost–benefit analysis has demonstrated that the action could be justified, and that specific objectives or targets are based on sound cost–benefit and scientific analyses (proposal 9). Any new proposal should be accompanied by a careful analysis of whether or not marked-based methods could be employed to achieve the same goals; where a market-based approach is feasible, any departures from it should be justified (proposal 10).

Both reports stress that implementation and enforcement of environmental legislation by the Member States should be strengthened. A comparative annual report should be developed and published by the Commission (Molitor Report: proposal 6).

NEW INSTRUMENTS AND A SHIFT IN RESPONSIBILITIES

However, reforming existing and future environmental legislation will not be sufficient for effective targeted reregulation and simplification. The traditional command-and-control approach has shown its limitations. That means that complements and/or alternatives need to be designed.

Proposal 5 of the Molitor Report recommends that as environmental policy increasingly shifts responsibility for implementation to the private sector, governments need to develop new ways to check that firms are meeting their obligations. The UNICE Report recommends that strategies for government intervention should give maximum rein to market forces, wherever possible, and emphasise the use of alternatives (such as negotiated agreements) to 'command-and-control' regulations.

Already in 1992, the EU's 5th Environmental Action Programme (EAP) 'Towards Sustainability' (European Commission, 1992) stated that a new sense of direction and thrust should be given to the environment/industrial policy interface by the institution of a comprehensive, integrated 'package' of measures, including existing provisions, comprising the following elements:

- a strengthening of the dialogue with industry;
- improved management and control of production processes including a system of licensing linked to integrated pollution prevention and control, environmental auditing, effective environment valuation and accounting, use of the best available technology, and introduction of market-based pricing systems for consumption and use of natural resources;
- higher, more reliable product standards designed to ensure that the environmental impact of products during their whole life-cycle is minimised;
- encouragement of voluntary agreements and other forms of self-regulation;
- effective waste management ideally should commence with the control of production processes.

According to the 5th EAP, it is essential that the general public and the social partners are enabled to become more actively involved in the development and practical implementation of policy, which means access to environmental information. An environmentally sound industry is no longer a matter of luxury but rather a matter of necessity, taking into account the increasing concern for the environment and natural resources, and realisation of the negative economic effects of environmental degradation.

It is equally clear that many sectors of industry are themselves becoming more appreciative of their relationship with and responsibility for the environment and the natural resource base. Therefore one of the key messages of the Programme is that in the field of the environment, industry must not only be part of the problem but also part of the solution (European Commission, 1992). European industry recognises the importance of a good environmental policy and has taken up what it sees as its own responsibility further to ensure the realisation of this policy. The following examples of initiatives within industry illustrate the way this responsibility can be taken.

The Business Charter for Sustainable Development

In its report 'Our Common Future', the World Commission on Environment and Development (Brundtland Commission) emphasised the importance of environmental protection in the pursuit of sustainable development (World Commission on Environment and Development, 1987). Sustainable development involves meeting the needs of the present generations without compromising the ability of future generations to meet their own needs. In April 1991 the International Chamber of Commerce (1991) introduced the Business Charter for Sustainable Development to help business around the world improve its environmental performance. It comprises sixteen principles for environmental management which, for business, is a vitally important aspect of sustainable development.

The first principle of the Charter is to recognise environmental management as among the highest corporate priorities and as a key determinant to sustainable development; to establish policies, programmes and practices for conducting operations in an environmentally sound manner.

There should be a common goal, not a conflict, between economic development and environmental protection. Economic growth provides the conditions in which protection of the environment can best be achieved, and environmental protection, in balance with other human goals, is necessary to achieve growth that is sustainable. In turn, versatile, dynamic, responsive and profitable businesses are required as the driving force for sustainable economic development and for providing managerial, technical and financial resources to contribute to the resolution of environmental challenges. Now, just a few years later, the Charter has assisted many companies in fulfilling their commitment to environmental stewardship in a comprehensive fashion.

British Standard 7750 (BS 7750)

Not only in the United Kingdom, but in other countries as well, the British Standard 7750 is used to guarantee that an environmental management system meets certain standards. Elements of BS 7750 are:

- an environmental management system;
- an environmental policy;
- organisation and personnel;
- personnel, communication and training;
- environmental objectives and targets;
- an environmental management programme;
- environmental management manual and documentation;
- operational control;

- environmental management records;
- environmental management audits;
- environmental management reviews.

The EU's Eco-Audit Scheme

In 1995, Council Regulation No. 1836/93 allowing voluntary participation by companies in the industrial sector in a Community Eco-Management and Audit Scheme (EMAS) came into force. Companies which volunteer to participate in this scheme undertake to establish an 'environmental protection system' for a given production site and to conduct a systematic, periodic evaluation of their environmental performance. They also undertake to provide the public with relevant, objective information on their performances and intentions as regards the environment. The protection system is based on an initial review, by the company, of the environmental impact of the operations conducted at the site. It includes an internal policy formally stated in writing, supplemented by a programme of measures at the site concerned and a management system covering the organisational details and working procedures needed to apply the above mentioned elements.

A key element in this management system is the environmental audit. This is the tool used by the company periodically to check that its management system is functioning properly. The credibility of the company's measures and commitments is ensured by periodically providing the public with objective, relevant information in the form of an environmental statement. This document describes the company's performance and intentions and is formally validated by an accredited independent auditor on the basis, inter alia, of the internal audit and its results. In return for satisfying the conditions for participation in the scheme, the scheme offers participating companies the possibility of capitalising on their commitment and boosting their public image through the use of a logo. An increasing number of companies is participating in the eco-audit scheme or is planning to do so in the near future.

Environmental reporting

In 1992, the World Industry Council for the Environment (WICE) published a brochure *Environmental Reporting. A Manager's Guide*. WICE is a global coalition of enterprises initiated in 1993 by ICC. The work programme of WICE includes research and policy recommendations reflecting concern for high standards of environmental management and commitment to the principles of sustainable development. In the above-mentioned brochure, WICE has attempted to provide sound, responsible suggestions on environmental reporting. WICE emphasises that it is up to individual enterprises to determine which of these suggestions are in their best interest. WICE believes that progress is best made voluntarily, with enterprises being free to decide what is most suitable for their needs, rather than requirements being imposed by regulation.

SUSTAINABLE INDUSTRIAL DEVELOPMENT (SID)

We saw that both governments and companies recognise today that environmental protection and economic development must go hand in hand. In fact this is what Brundtland's Sustainable Development is all about. However, the problem is that this concept of Sustainable Development is not suitable for use as such in business practice.

In February 1996 the Dutch Ministers of Housing, Spatial Planning and the Environment, and Economic Affairs, hosted a conference 'Sharing Responsibilities in a Competitive World' where the concept of Sustainable Industrial Development (SID) was elaborated. The fundamental idea of SID implies that two objectives are achieved simultaneously:

- strengthening the position of business in a competitive world; and,
- a sustainable use of the world's natural resources.

In the framework of this conference, UNICE attempted to give a business content to the concept of SID. For business, the concept of SID has four main components:

- *Integration*. The need to integrate environmental considerations in day-to-day business management. From plant operation to purchasing policies, from office management to waste disposal.
- *Anticipation*. The capacity of companies to anticipate environmental opportunities, constraints and conditions.
- *Operational applications*. Even though SID is more a background concept and determinant for management decisions, it can and must find direct, operational applications as well.
- *Dissemination*. It is crucial that not only large companies, but small and medium-sized enterprises as well subscribe to SID principles.

Let us elaborate on each of those dimensions of SID. Integration is often easier said than done. Environmental considerations are far from being the sole determinant of business management. A company's main drive is the benefit it can yield from its operations. Profitability may suffer from the integration of environmental concerns, as can competitiveness and, in some cases, employment. But, on the whole, the development of an environmental management system within a company is likely to deliver beneficial effects. For instance by analysing the way some processes are impacting on the environment and correcting problems, significant savings can be achieved, for example in terms of energy or raw materials consumption.

Anticipation requires a long-term view, and is mainly a matter of asking oneself the right questions for strategic decisions on the 5- to 10-year horizon. A long-term view means identifying and evaluating medium- to long-term environmental issues and trends. It means also collecting signals about these trends in terms of public perception and likely government policies. After gathering the information, a company or even a sector needs to design responses to these trends, be it in terms of business policy and planning strategy, investment decisions or process and product design.

At operational level, SID can find more direct applications in the idea of ensuring eco-efficiency for products and processes. In this dimension, SID has two equally important components: an ecological and an economic one. The ecological component asks for a reduced and/or more efficient use of natural resources, especially non-renewables, and strives towards minimum environmental impact of industrial activities. The economic component means that environmental measures must take into account the economic efficiency of actions taken and policies made.

Dissemination is needed for small and medium-sized enterprises (SMEs). As mentioned in the recently adopted progress report on the implementation of the EU's Fifth Environmental Action Programme (European Commission, 1995b), large companies have now by and large integrated the environmental dimension. For instance approximately 30 major European chemical companies are issuing detailed environmental reports concerning their activities on the basis of European guidelines, adopted by CEFIC, the European Federation of Chemical Industries. However, SMEs represent a much bigger part of employment and overall investment in the European Union. Therefore, it is crucial that SMEs subscribe to the principles of SID.

SHARED RESPONSIBILITY FOR IMPLEMENTING SID

The four components just described, represent the basic industrial dimension of SID. While there is still confusion and considerable debate about exactly what SID means, it seems more adequate to make SID a fully operational tool than only to have the debate going on about the concept itself. SID needs action rather than reaction from the industry sectors concerned.

As pointed out in the paper prepared by A.D. Little for the Dutch Conference on SID, it is generally recognised that SID needs new ways of thinking and working, and these need to draw upon recent initiatives where collaboration between industry and government is promoting changes. This includes that environmental policy for industry becomes less focused on regulation and control, and more on cooperation and incentives, and business management focused as much on the competitive use of environmental resources as on other business imperatives.

174

Or, as Pieter Winsemius, Director of McKinsey and Company and former Dutch Minister of the Environment, puts it:

> Governments are no longer able to prescribe optimal solutions to environmental problems . . . Even environmental organisations have come to realise their own limitations . . . Today, only true industry experts have the understanding of the often intricate process of product design to come up with solutions that secure a desired environmental improvement at a minimal cost, and in such a way that the burden is shared fairly among all polluters.
>
> (WICE, 1994)

So what really is necessary is the use of new instruments, based on cooperation between government and industry. This fits with the above-mentioned developments and leads to the conclusion that now is the time to establish a framework for this cooperation. A shared agenda will be a precondition for a fruitful cooperation between government and industry. This agenda should be based on the five principles of engagement A.D. Little mentions:

- *A shared vision* of SID, based on agreement about the challenges (quantum improvements) and underlying principles: both sustainable environmental quality and sustainable competitiveness.
- *Commitment to progress toward the shared vision*, demonstrated through definition of explicit objectives and targets, and allocation of responsibilities and resources for both short- and long-term actions.
- *Transparency of progress and performance* against objectives and targets based upon open reporting, accountability, and commonly accepted measures and metrics.
- *Trust* as a precondition to collaboration based on the key processes of dialogue, negotiation and consensus.
- *Creativity* in identifying, developing and adapting innovative technological, policy, and management solutions.

At present, no one country has the key to SID. But, as we will see hereafter, many countries have experience of different kinds of partnership between governments and companies. This experience can be used as reference for similar situations and is a good way to demonstrate that partnership can deliver effective results and environmental improvements.

However, partnership should not create a confusion of roles between industry and government. Industry views partnership as a means to make its technical know-how and experience available to public authorities. Partnership is not a substitute for the political decision-making role of government and parliament that is the basis of our democracies. Partnership is a facilitating process towards the achievement of sustainable objectives.

175

SID requires the application of policy instruments beyond the traditional 'command-and-control' regulations. Once the objectives are clear, new instruments must be designed in partnership, such as economic instruments, voluntary approaches by industry or negotiated agreements between public authorities and industry. Negotiated agreements and other voluntary approaches should be used because of the positive experience of both governments and companies with these methods.

NEGOTIATED AGREEMENTS

For the purpose of this paper, negotiated agreements are understood as meaning any commitment with the aim of protecting or restoring the environment, undertaken by individual companies and/or associations of companies, which is the result of negotiations with the competent public authorities and/or which are explicitly recognised by these authorities. The term 'negotiated agreement' is preferred to 'voluntary agreement', because of two main reasons. Firstly, it is very important to stress that an agreement is the result of negotiations. Secondly, the term voluntary could lead to the false conclusion that companies have a fully open choice to agree. That is not the reality, because usually the choice will be between an agreement or legislation.

Negotiated agreements could be concluded at European level between the European institutions and (associations of) companies (European agreements). Negotiated agreements concluded at national, regional or local level, can be divided into implementation agreements, which implement Community Directives, and national agreements, which are not meant to implement Community legislation. In 1996, there was already a large number of voluntary actions and national agreements in most, if not all, Member States of the EU.

In 1995, the International Energy Agency (IEA) presented the background paper 'Voluntary Actions for Energy-related CO_2 Abatement in IEA Member Countries', which mentions voluntary actions in Australia, Austria, Belgium, Canada, Denmark, the European Union, Finland, France, Germany,[3] Greece, Ireland, Italy, Japan, the Netherlands, New Zealand, Norway, Portugal, Spain, Sweden, Switzerland, Turkey, the United Kingdom and the USA.

UNICE estimates that more than 250 negotiated agreements, valid for between 5 to 10 years, exist in Member States, particularly in the Netherlands, Germany, and Austria, and have already produced significant results.[4] Non-binding declarations such as 'Responsible Care' or 'gentlemen's agreements' are not considered to be negotiated agreements.

The 5th EAP, the Communication on 'Industrial Competitiveness and Environmental Protection' and the Council resolutions linked to these policy documents, refer to negotiated agreements as a possible means for promoting environmentally sustainable and pro-active practices in industry, and as a way to fill in 'shared responsibility'. This recommendation was picked up and restated in the so-called Delors White Paper on Growth, Competitiveness and Employment (European Commission, 1993), and was made one of the principal subjects for discussion at the follow-up conference on the White Paper conclusions later that year (Delors II conference).

UNICE has responded positively to these policy statements from the Community. It organised a high-level workshop on 28 March 1995 with the participation of inter alia the Commissioner for the Environment, Members of the European Parliament and representatives from European Industry and Industry Federations. On 6 March 1996 the EC informed the Member States about a proposal for a Communication on negotiated agreements, to be presented in mid-1996. However, at the time of writing, this had not been published.

POTENTIAL BENEFITS OF NEGOTIATED AGREEMENTS AND HOW TO MAKE THEM WORK

The main reasons cited in the above mentioned IEA survey for adopting voluntary actions were:

- to promote increased involvement of firms in emissions reduction activities;
- to increase industries' and consumers' motivation and responsibilities to achieve environmental objectives; and
- to permit the definition of policy instruments better adapted to the economic and competitive context.

These are important benefits for industry. For governments, other potential benefits could lie in the use of negotiated agreements, such as:

- the encouragement of a pro-active industry;
- faster achievement of objectives as compared to sometimes lengthier adoption of legislation;
- 'stop-gap' where knowledge is insufficient for the legislator to intervene.

To make a voluntary approach successful, we assume that at least four general conditions have to be met:

- *Clear and achievable targets.* There is no point in asking groups in society, including industry, voluntarily to reduce their environmental pollution if it is not clear what kind of reduction is required or if the reduction being demanded is even technically or financially achievable. The need for this reduction is to be understood and accepted.

177

- *Recognition of mutual benefits.* Both government and industry will need to be convinced of the advantages of working together. This will require a considerable level of conviction and information, both within government and within industry.
- *The threat of punitive measures.* Cooperation based on consultation and consensus cannot and must not mean that the government does not have recourse to statutory regulations, licences and sanctions which can be applied to those who do not comply with the rules or licensing requirements. Threats are also helpful to industry in serving to limit free-rider behaviour and the distortion of competition.
- *Trust.* In fact this is the key to success. The whole voluntary approach is built on trust from both sides. This requires that each party's responsibilities are clearly defined and properly respected and that targets are mutually agreed. But, of course, trust has to be earned.

CONTENTS OF NEGOTIATED AGREEMENTS

Experience teaches us that it is not possible to provide a model for making a negotiated agreement. However, it is possible to give some guidelines, for example based on the Dutch experience. Most important is that the contents shall ensure credibility, reliability and transparency of the agreement. Key issues are:

- the signatories must be clearly described as well as the tasks and responsibilities of the parties. They must have the power to fulfil the obligations set in the agreement;
- clear and achievable objectives;
- in order to make the agreement credible, appropriate monitoring mechanisms should be in place;
- in order to make the agreement transparent, the results achieved should be made available to the public. In addition, regular reporting on the execution of the agreement is important to assess compliance with the agreement;
- revision of the agreement should be foreseen under specific circumstances, e.g. substantial technical progress or economic circumstances;
- compliance with the EC Treaty and in particular with internal market and competition rules.

There are a number of other issues, which have to be considered when making a negotiated agreement, such as the nature of obligations, how to enforce the agreement, what sanctions to apply for non-compliance, how to deal with free-riders, the relation of the agreement to permits and legislation, how long the agreement is to last, and what access to information needs to be made available.

178

CONCLUSIONS

We believe that negotiated agreements can be considered as an instrument in the search for the most effective 'instrument mix'. It is not meant as a substitute for regulation. The use of negotiated agreements is not and cannot be an end in itself. What matters is to work in partnership and:

- to define clear, reliable and feasible objectives;
- to design the most appropriate policy instrument to meet them, which presupposes an analysis of environmental efficiency and economic impact, especially in terms of competitiveness of industry;
- to define roles and responsibilities in establishing a performance progress.

Applying SID and developing negotiated agreements is not a matter just for individual companies or public authorities. Associations and federations of companies should play an important role to echo companies' preoccupations and ideas at EU level and to focus the EU institutions on priority issues and actions, taking economic and more specifically competitiveness constraints into account. Furthermore, when necessary, they should act as coordinator for industry input on specific SID issues and negotiated agreements. The European institutions (the Council, the Commission and the European Parliament) should play an important role as well in developing and supporting negotiated agreements.

NOTES

1 The views expressed in this chapter are the views of the author in his capacity of Chairman of UNICE's Environment Working Group.
2 An example would be the Directive on Integrated Pollution Prevention and Control (IPPC), proposed by the Commission in 1993 and recently adopted by the Council. On the basis of IPPC, Member States have to take the necessary measures to ensure that no installation is operated without a permit in order to achieve a high level of protection for the environment as a whole. The conditions of the permit must normally include at least emission limit values for a list of substances and preparations. These emission limit values (ELVs) must be based on the Best Available Techniques (BATs).
3 For instance the one-sided Declaration by German Industry and Trade on Global Warming Prevention on 10 March 1995.
4 Press Release UNICE on 28 March 1995.

REFERENCES

European Commission (1992) 'Towards sustainability', *COM* (92) 23 final.
European Commission (1993) 'Growth, competitiveness and employment: the challenges and ways forward into the 21st Century', *COM* (93) 700 final.
European Commission (1995a) 'Report of the group of independent experts on legislative and administrative simplification', *COM* (95) 288 final.

European Commission (1995b) 'Review of the Fifth Environmental Action Programme', *COM* (95) 624.

International Chamber of Commerce (1991) *The Business Charter for Sustainable Development, Principles for Environmental Management*, Publication 210/356 A, Paris: International Chamber of Commerce.

UNICE (1995) *Realising Europe's Potential through Targeted Regulatory Reform: the UNICE Regulatory Report*, Brussels: UNICE.

WICE (1994) *Improving Policy Cooperation between Governments and Industry*, Report of the WICE Working Group on Policy Partnerships, Geneva: International Chamber of Commerce.

World Commission on Environment and Development (1987) *Our Common Future*, Oxford: Oxford University Press.

11

VOLUNTARY AGREEMENTS AS A FORM OF DEREGULATION?

The Dutch experience

Duncan Liefferink and Arthur P.J. Mol

INTRODUCTION

So-called voluntary agreements between the state and private actors, particularly industry, are in fact seldom entirely voluntary. Quite often, they are linked to more general legal obligations and can as such rather be seen as implementation agreements. And even if they are not placed in a broader legal context, the state may use the introduction of formal regulations as a stick to beat with if 'voluntary' negotiations do not bring the desired results. Nevertheless, voluntary agreements are often regarded as a form of deregulation, as they are assumed to reduce the regulatory burden on industry. However, examining the phenomenon of voluntary agreements more closely this view may be questioned. Although the use of voluntary agreements, or more interactive modes of policy-making generally, may indeed give room to firms for more flexible responses to environmental requirements, focusing on this aspect appears to be somewhat one-sided. In this paper it will be argued that the increasing popularity of voluntary agreements should rather be seen as an expression of the shifting role of the state in dealing with environmental problems in a more encompassing sense. This shift can be associated with the crisis of the welfare state which gradually took shape during the 1970s and 1980s, and, in the environmental field, with the perceived ineffectiveness of traditional 'command-and-control' strategies.

In the next section the theory of ecological modernisation will be used to explore the shift to a new phase of environmental policy-making, characterised, among other things, by the increasing importance of interactive and participative approaches. Ecological modernisation can be seen as a substantial theory on a high abstraction level. One of its focuses is the changing relationship between the state and polluters, following upon the debates on market failure and state failure in traditional environmental policy.

In the three following sections we will examine in considerable detail the shifting relations between public and private actors, and more specifically, the rise of voluntary agreements in the Netherlands. Compared with other countries, the Netherlands already has a rich experience with interactive policy approaches. After a somewhat isolated and largely unsuccessful deregulation effort in the environmental policy field in the early 1980s, a more encompassing re-orientation of Dutch environmental policy was carried through from the mid-1980s. Starting from the idea of 'internalisation' of environmental responsibilities by the 'target groups' of environmental policy, a more collaborative relationship between public and private actors was aimed at. Interactive policy-making and covenants played a central role in this approach. So far, more than 100 environmental covenants between the state and various industrial sectors have been concluded. This makes the Netherlands an excellent case for studying the central characteristics, as well as the theoretical implications of voluntary agreements. First, the history of what has come to be known as the target group approach will be analysed. After that, the practice of environmental covenants will be investigated with the help of a case study in the field of the dairy industry, followed by an evaluation of the pros and cons of an interactive approach in environmental policy.

In the concluding section we will attempt to pin down more precisely the nature of the observed shifts in the relationship between the state and polluters. In particular, we will discuss to what extent the development of the Dutch target group approach and the ideas it incorporates may indeed be interpreted in terms of ecological modernisation and how this can be related to the presumed deregulatory aspects of voluntary agreements.

CORE FEATURES OF ECOLOGICAL MODERNISATION THEORY

The concept of ecological modernisation is gaining increasing popularity. Particularly the ideological vanguard of some Dutch and German environmental organisations and political parties has made significant contributions to the definition and promotion of ecological modernisation as a *political programme* for the reform of environmental politics (see for instance Schöne, 1987; Fischer, 1991; van Driel *et al.*, 1993). The political programme of ecological modernisation has also been described, analysed and further elaborated by authors like Simonis (1989), Weale (1992) and Andersen (1994). In this chapter, however, we will mainly refer to ecological modernisation as a *theory* of social continuity and change (cf. Huber, 1982, 1991; Zimmermann *et al.*, 1990; Spaargaren and Mol, 1992; Jänicke, 1993; Hajer, 1993; Mol, 1995).

Ecological modernisation as a theory basically deals with the institutions of modern technology, (market) economy and state intervention in a long-term perspective. The theory focuses on the ecological transformation of production and consumption, which is currently taking place in Western Europe and can be regarded as the third and, until further notice, final phase in the historical development of modern, industrial society. The preceding phases were the industrial breakthrough or industrial take-off (1789/1793 until 1848) and the construction of industrial society (1848–1980). The central notions of ecological modernisation theory can be summarised in three points which distinguish it from other social theories of environmental reform, such as the theory of risk society, post-modernist approaches, neo-Marxist and counter-productivity theories (see further Mol, 1995: 7–26). Particularly the third point, discussing the role of the state in the process of ecological modernisation, directly relates to the phenomenon of voluntary agreements in environmental policy.

First, and in line with the above Schumpeterian view on the development of industrial society, ecological modernisation theory identifies modern science and technology as principal institutions for ecological reform, rather than as the main culprits of ecological and social disruption. Science and technology are central to what may be called the process of 'ecologising economy'. In the era of reflexive modernity and in confrontation with the ecological crisis, scientific and technological trajectories are changing. The simple end-of-pipe technologies that have been criticised so strongly in the 1970s (e.g. Jänicke, 1979) are increasingly replaced by more advanced environmental technologies that not only redirect production processes and products into more environmentally sound ones, but also start to set about the selective contraction of large technological systems that can no longer fulfil stringent ecological requirements. In that way, the role of technological measures within ecological modernisation theory is far beyond that of providing 'technological-fix' solutions.

Second, ecological modernisation theory stresses the increasing importance of economic and market dynamics in ecological reform and the role of innovators, entrepreneurs and other economic agents as social carriers of ecological restructuring, in addition to state agencies and new social movements. In its rejection of the fundamental opposition between economy and ecology, it is akin to the Brundtland concept of sustainable development (WCED, 1987). Modern economic institutions and mechanisms can be – and are to an increasing extent – reformed according to criteria of ecological rationality. The internalisation of external effects via 'economising ecology' is one of the mechanisms that can be observed in this context (cf. Andersen, 1994), as well as the articulation of environmental 'standards' in economic processes, for instance by insurance companies, credit institutions, (industrial) consumers, certification organisations and branch associations.

The third core feature of ecological modernisation theory addresses the role of the state. Following the discussions on state failure in, among others, environmental policy (cf. Jänicke, 1986), ecological modernisation is critical

183

of the traditional, strong and bureaucratic state in the redirection of processes of production and consumption. This is not to say, however, that the state has become totally irrelevant to the management of environmental problems. Rather, its role in environmental policy is observed to be changing. As the old curative and reactive policies increasingly turn into preventive ones, polluters are more and more directly involved in the formulation and implementation of policy measures. Hierarchical and democratically legitimated intervention is being replaced by dialogue and contextual steering. Problems are discussed and solved at a more decentralised level with the state in a facilitating rather than an interventionist role. This goes together with a increasing reliance on unorthodox, 'non-regulatory' forms of (often mutual) commitment between public and private actors. From this point of view, it is not surprising that certain tasks, responsibilities and incentives for environmental restructuring are entirely shifting from the state to the market. As already referred to above, private economic actors become involved in environmental reform for instance by certification of products and processes, by asking for environmental audits, by competition on the basis of environmental performance and by the creation of niche markets. By leaving fewer (but essential) elements of environmental policy-making for the central state, and by changing the interrelation between state and society/economy, the state is prevented from becoming an environmental Leviathan (cf. Paehlke and Torgerson, 1990).

The shifting role of the state in environmental policy-making has been most strongly emphasised by Jänicke (1993), who speaks of a process of political modernisation parallel and complementary to that of ecological modernisation. Others have referred to the same tendencies with the concept of reflexive steering (Le Blansch, 1996). Particularly this aspect of ecological modernisation theory may help to understand the present rise of interactive and participative approaches in environmental policy-making. In the following sections the Dutch experience in this field will be examined. To what extent can the early emphasis on deregulation arguments be traced back also in the later, more encompassing re-orientation of Dutch environmental policy? Or should the widespread use of voluntary agreements in the Netherlands indeed primarily be interpreted as a step in the process of ecological modernisation? In order to answer these questions, both the ideological debate around the emergence and development of the 'target group' approach and process of making and implementing environmental covenants will be analysed.

INTERNALISATION AND VOLUNTARY AGREEMENTS: A HISTORICAL ANALYSIS

As in most other industrialised countries, the first phase of environmental policy in the Netherlands had been focused on cleaning up the most threatening problems with the help mainly of end-of-pipe technology.[1] Various laws had been adopted dealing for instance with water pollution, air pollution, waste and noise. Although a remarkably effective system of effluent levies had been installed in the field of surface water pollution in 1970 (cf. Bressers, 1983; Andersen, 1994; Liefferink, 1997b), policies in this period relied to a large extent on direct regulation. General standards were set at the national level and formed the basis of licences for individual plants or activities.

Around 1980, the inadequacy of this type of policy became increasingly apparent. While it had been estimated in 1971 that the most serious problems could be solved within five to ten years (*Urgentienota Milieuhygiëne*, 1971–1972), a further deterioration of the environment was observed in almost all respects. Additionally, the role of the state in a more general sense was discussed more and more openly. The economy stagnated and budgetary problems were experienced by the government in The Hague. For the first time since its inception, there was talk about a crisis of the welfare state. Although more fundamental attacks on the institutions at the very centre of the welfare state, notably the social security system, were as yet to come in the 1990s, industry argued for a 'leaner and meaner' government. Particularly the heavy and highly obscure regulatory burden imposed on firms was seen as a major barrier to new economic initiatives. Thus, deregulation became one of the key-words of the Christian-Democrat/liberal coalition which took over in 1982 (in more detail: Hanf, 1989).

Not surprisingly, environmental policy was one of the first fields in which deregulation was attempted. The large variety of environmental laws, each with their own licensing authorities and procedures, had for a long time been a thorn in the flesh of industry. A start with integrating the many procedures had already been made with the General Environmental Provisions Act of 1980. In 1983, a more encompassing action programme on deregulation in the fields of physical planning and environmental management was published (Aktieprogramma DROM, 1982–1983). The document only evaluated the consequences of regulation for industry, but did not call into question the government's environmental policy goals as such. With this limitation, the programme actually contained only a few substantive proposals, notably the development of standard licences for certain categories of small polluters and the introduction of regional 'bubbles' for air pollution. The former suggestion was indeed soon put into practice, whereas the 'bubble' concept is still under study and discussion. Apart from that, an impressive number of minor recommendations was made with regard to the implementation of various existing regulations. At best, in conclusion, the operation led to the replacement of rules by other rules, or in other words, to reregulation (cf. van Nispen and van der Tak, 1984; Gastkemper and Klijn, 1986).

The limited effect of the action programme on environmental deregulation once more demonstrated that streamlining existing regulation was not enough. In order to raise the economic efficiency of environmental policy, more fundamental changes were required. This became even clearer when policy effectiveness was considered. Decisive steps towards a far-reaching re-organisation of Dutch environmental policy were taken in the mid-1980s. The main objective of these efforts was to achieve a better integration both of the until then fragmented sectoral elements of environmental policy itself, and between environmental policy and other policy sectors.

In successive 'Indicative Multi-year Programmes' for the Environment (IMPs), published annually from 1984 onwards, two complementary policy lines were set out. In the first place, 'effect-oriented' policies were organised around a number of major environmental problems such as acidification, eutrophication, and waste, cross-linking the traditional 'compartments' (air, water, etc.). Second, coherent 'source-oriented' policies were to be developed for specific target groups of environmental policy, such as industry and refineries, traffic and transport, and households (or consumers). An important objective of the 'source-oriented' line was to bring about a process of 'internalisation' of environmental responsibility with the target groups. The concept of internalisation as introduced in the mid-1980s (cf. particularly IMP-M 1985–1989; IMP-M 1986–1990) is indeed a major key to understanding the discussion about regulation, state–industry relations and voluntary agreements in Dutch environmental policy.

Internalisation has two different aspects. On the one hand, it refers to the need to raise the willingness of polluters to contribute to the solution of environmental problems by invoking their sense of 'social responsibility'. On the other hand, as a kind of counter-effort on the part of the government, it entails paying serious attention to the needs and wishes of groups affected by environmental policies and taking them into account in the design of measures. The new policy approach was instigated primarily by Pieter Winsemius, a member of the liberal party and Minister of the Environment from 1982 to1986. The discussion about internalisation as it was carried on in the Netherlands had slightly moralistic overtones. This was aptly summarised by the appeal to behave like a 'Guest in one's own house', the title of Winsemius's political testament (Winsemius, 1986). At the same time, however, a close link with the central concepts both of the Brundtland Report (WCED, 1987) and of ecological modernisation can be recognised, particularly in the strong belief in the mutual advantages of a relationship between state and industry based on equality and synergy, or so-called 'win-win' situations (see also Le Blansch, 1996). Not surprisingly, the Brundtland Report, which was in fact published somewhat later and strongly emphasised the partnership between economy and ecology, was easily adopted as an additional basis for the prevailing trends in Dutch environmental policy in the late 1980s. Also reminiscent of the ideas of ecological modernisation is the preference for a shift of certain operational responsibilities to the target groups as a way to stimulate internalisation. Such shifts, Winsemius argued, could be laid down in voluntary agreements between the government and industrial sectors (Winsemius, 1986).

A further feature often associated with the internalisation approach is the diversification or shift of policy instruments from traditional direct regulation to the use of different kinds of incentives and communication. It was convincingly pointed out by Le Blansch, however, that this is difficult to maintain empirically. Also the old command-and-control strategies, particularly in the consensus-oriented democracies in the North-western part of Europe, often have an element of negotiation with the target group in the early phases of decision-making. It is by no means an exception, on the other hand, that covenants are directly linked to 'harder' policy instruments. Covenants may serve as a management tool for implementing more general obligations laid down by law. This was for instance the case when Dutch oil companies agreed to set up a fund for the clean-up of contaminated sites of petrol stations (VROM, 1994). It also occurs that agreements originally made on the basis of civil law are later, as it were, sealed under public law. Policy instruments have thus always been and will always be mixed, but the distinctive feature of the internalisation approach is that the choice of the mix in each particular situation is more than before subject to reflection and negotiation (Le Blansch, 1996: 45).

Policies based on internalisation, in short, can be regarded as a common learning process, in which both means and objectives are established by public and private actors together. The process of reflection, negotiation and agreement is not only a way to formulate concrete policy measures, but it is also a goal in itself, contributing to the internalisation of environmental responsibility by the target groups.

ENVIRONMENTAL COVENANTS AND TARGET GROUPS: PRACTICAL EXPERIENCES

As the foregoing sections have demonstrated, voluntary agreements do not stand alone but are part of a larger concept of government intervention. This concept is characterised by close interaction between public and private actors, and reflection about both means and goals. Especially in the Dutch case, it is intimately linked with the idea of the internalisation of environmental responsibility. Without going into too much detail, this section seeks to explore the process of making and executing a covenant in the framework of the Dutch target group approach by taking the dairy industry as an example.[2] In Dutch environmental policy-making two types of covenants can be distinguished:

- sectoral covenants that cover a wide range of environmental issues within one economic sector;
- covenants that focus on a particular environmental issue, such as energy saving, packaging or cadmium.

Before focusing on the case of the dairy industry, it may be useful to outline the systematic approach that has been developed, especially for the former type of covenants.

Sectoral covenants

The process towards sectoral covenants in the Netherlands follows a more or less standardised procedure (VROM, 1994; in more detail: VROM, 1992). The initial steps are similar for all target groups. In the second half of the process, a distinction is made between homogeneous and heterogeneous sectors. In the first phase, carried out in the early 1990s, broad environmental objectives as laid down for instance in the National Environmental Policy Plan (NEPP, 1988–1989) were elaborated for specific sectors. This was done primarily by the Ministry of Housing, Spatial Planning and the Environment (*Ministerie van Volkshuisvesting, Ruimtelijke Ordening en Milieubeheer*, VROM). Detailed emission inventories provided an overview of each sector's contribution to the main environmental problems or 'themes', such as acidification and climate change. On the basis of these inventories, a number of priority branches was selected for the target group approach. This selection included fifteen major industrial branches[3] and covered over 90 per cent of total industrial pollution in the Netherlands. For each of these branches, basic long-term targets for emission reductions were established. Sectors that were not selected continued to be regulated with the help of traditional licences.

Already during the first phase, preliminary talks were held with the target groups, but the actual consultations are taking place in the second phase. With the exception of the long-term reductions targets,[4] all aspects of the measures are open for discussion in this phase, including time paths, methods of implementation and monitoring. Apart from the Ministry of VROM and dependent on the sector involved, other ministries (e.g. Economic Affairs, Transport and Public Works) take part in the negotiations, as well as provincial and municipal authorities. The latter are involved particularly because of their role in granting licences. Industry is represented by the relevant branch organisation. In several cases, this has led to the revitalisation of old, inactive organisations (e.g. the Dutch Foundation Packaging and Environment, cf. Mol, 1995: 238–243) or the establishment of new ones solely for this purpose (e.g. the Foundation for the Basic Metals Industry and the Environment, cf. VROM, 1994: 4). Trade unions may be represented, but hardly ever are in practice. Environmental organisations are sometimes consulted in the earlier stages of the negotiations, but are usually not among the core negotiating community, nor among the signatories of the agreement.

The conclusion of the covenant at the sector level takes place in the third phase. It forms the basis for practical implementation plans. In homogeneous sectors, characterised by the application of a limited range of processes and technologies, practical measures are now worked out also at the branch level. They specify for instance measures to be taken within a certain time limit, as well as organisational and procedural arrangements for the communication of the commitments to the individual companies, and for monitoring and enforcement. Implementation plans at the branch level can again take the form of a covenant under civil law.

In heterogeneous sectors, i.e. the majority of the initial fifteen branches, implementation plans are elaborated at the level of individual firms. It is drawn up by the firm itself in close collaboration with the relevant licensing authority and forms the basis for issuing permits. Due to this link with the traditional licensing procedure, the licensing authorities bear the initial responsibility for the enforcement at the firm level. In addition, a system of feedback to the sector level and the Ministry of VROM is warranted.

Dairy industry

The dairy industry was identified by the Ministry of VROM as one of the sectors in urgent need of a sectoral environmental policy in view of the heavy burden it placed on the environment. The most serious environmental problems caused by the dairy industry are related to waste water discharges, air emissions related to high energy use, dust, odour, solid waste and a large water consumption. With some 15,000 employees in 25 companies and 95 production sites, and a diversity of products the dairy industry is seen as a heterogeneous sector. It is dominated by cooperatives, which process 84 per cent of total milk production, and is concentrated in the eastern and northern parts of the country.

Negotiations for a sectoral covenant started in 1992 and resulted in the publication of a Declaration of Intent in June 1994. By then, some 16 meetings had been held, newsletters had been produced and communication between the participants and the individual companies was frequent. With the exception of CO_2, which was dealt with in a separate covenant (see below), the covenant contained environmental objectives regarding all major environmental issues, usually in terms of emission reduction targets of some 30–50 per cent by 2000 and 2010 relative to 1985. The reduction targets negotiated basically followed those set out in the 1989 National Environmental Policy Plan.

A second phase started in 1994 with drawing up so-called Company Environmental Plans (CEPs) by each company. These plans listed three types of measures and related reductions:

- guaranteed reductions, related to measures which can be taken without further discussion;
- conditional reductions, following from measures that can only be taken if certain conditions are fulfilled (e.g. feasibility related to technology or the international economic situation);
- uncertain reductions, related to uncertainties that first need to be removed before a final judgement can be given on the applicability of the measures.

By April 1996, all CEPs were completed and sent to the competent authorities. During the negotiations on the Declaration, the dairy industry was represented by its two branch associations, the Netherlands' Dairy Organisation (*Nederlandse Zuivel Organisatie*, NZO) and Nedsmelt.[5] Two of the three companies which were neither a member of NZO nor of Nedsmelt completed a CEP, so that the problem of free-riders hardly existed. The 'other' side was formed by the Ministry of VROM and representatives from three other ministries, as well as local and provincial authorities and the regional water boards. Upon request from the dairy industry, no environmental or consumer organisation participated. The Ministry of VROM took the lead in the negotiations: it invited the other governmental participants, it set the agenda, objectives and time frame, and it dominated the discussions on the governmental side. The actual negotiation process was facilitated by an intermediate organisation called FO-industry.[6] In this phase it was agreed, among other things, that the covenant was to be seen as an additional instrument rather than as an alternative for permits, but it was agreed that environmental objectives of accepted CEPs were to be used in environmental permits (up to the year 2000).

During the process of implementing the Declaration and drawing up CEPs, a Consultative Group – and under it a Project Group – were set up, consisting of representatives of all organisations that participated in the first phase. Companies identified the following advantages of CEPs:

- the possibility of setting priorities and phasing environmental measures on a relatively long-term basis, thus matching environmental with economic investments;
- utilising the group approach to economically optimise emission reductions;
- concentrating contacts with public authorities regarding various environmental issues on one desk.

No third parties were involved in the process of making CEPs. If conflicts arose, for instance regarding the prioritising and phasing of measures and their formalisation into permits, legal resources were occasionally (and successfully) used by government officials.

Two other environmental covenants involving the dairy industry were signed in the early 1990s, one regarding energy and the other regarding packaging. The Long Term Agreement on Energy Efficiency Improvement is an agreement between the Ministry of Economic Affairs, responsible for energy policy, and NZO and Nedsmelt as representatives of the dairy industry. It was signed in July 1994 and lays down a 20 per cent improvement in energy efficiency by 2000 relative to 1989, as a contribution to the national 3–5 per cent reduction target for CO_2 by 2000.[7] A sectoral Energy Efficiency Plan with concrete measures was designed and subsequently elaborated in Energy Saving Plans (ESPs) at the level of individual industries. Due to sensitive and secret information, the details of these ESPs are confidential and not accessible for the government or for third parties. The independent consultancy organisation NOVEM (*Nederlandse Organisatie voor Energie en Milieu*) evaluates all ESPs and monitors their implementation. NOVEM in turn reports annually to the Ministry and is thus the key to public and political control of energy efficiency improvement. This is a major difference with the former covenant. The results of the agreement are encouraging for the sector: the intermediate goal of 8 per cent energy efficiency improvement by 1996 is in reach, and no major obstacles are expected with respect to the final objective in 2000. It should be noted, however, that the overall target of a 3 to 5 per cent reduction of CO_2 emissions until 2000 will not be achieved due to economic growth.

The Packaging covenant was the second major environmental covenant in the Netherlands (after the so-called 1989 'KWS-2000' covenant on volatile organic compounds). It was signed in 1991 between the Ministry of VROM and the Foundation Packaging and Environment (*Stichting Verpakking en Milieu*, SVM), being the representative of the packaging sector. Eight big dairy companies (accounting for 90 per cent of total milk production) are members of SVM. The covenant established general national targets of 10 per cent waste prevention and 55 per cent waste re-use/recycling (NEPP-Plus, 1989–1990) for the packaging sector, by specifying targets for prevention, product re-use, material re-use (recycling) and re-collection of packaging waste for different flows of packaging material and different years.

Although the organisational model for the negotiations (a facilitator and a limited number of representatives from private and governmental organisations) was comparable, the packaging covenant differs in a number of ways from the former two. As the packaging issue was of major political interest and since environmental and consumer organisations were directly involved during a significant part of the process, the negotiations were considerably less secluded and public and political debates interfered relatively strongly. Due to the diversity of the packaging chain and the

191

existence of a considerable free-rider problem (SVM represents only some 60 per cent of turnover in the packaging industry), the industrial front was less unified and negotiations accordingly complicated. Especially the parallel development in Germany of a formal decree on packaging enhanced the effectiveness of threatening with the legal 'stick' by the Ministry of VROM.

During the negotiations, the dairy industry hardly played a separate role, although one company (but not the branch association NZO!) was represented in the management board of SVM. Implementation of the covenant does not involve the regional and local authorities and no linkage is made with environmental permits; monitoring of the results takes place only at the level of sectors. This makes the introduction of measures by individual companies under this covenant more voluntary than measures under the two other covenants. In 1996, however, legal measures had to be taken as a consequence of the EU's Directive on Packaging and Packaging Waste.[8] As the maximum targets of the EU Directive are less ambitious than those of the covenant and, moreover, the Ministry of VROM had already taken up the idea of tightening certain objectives of the original covenant, the Ministry is now investigating the possibility of negotiating a new packaging covenant in addition to the legal decree. Although part of the Dutch industry is still loyal to the old covenant, however, little enthusiasm exists for going further than the targets set there.

ADVANTAGES AND DISADVANTAGES OF COVENANTS

As pointed out above, covenants have become extremely popular in Dutch environmental policy and all industrial sectors are confronted with this new instrument. The considerable advantages which may account for this include flexibility and speed, particularly in comparison with the lengthy process of the enactment of a formal law. Furthermore, the direct participation of polluters in policy formation is indeed widely claimed, not least by industry itself,[9] to raise their commitment to environmental policy goals. The approach would thus indeed contribute to the 'internalisation' of environmental responsibility. Related to this and in line with ecological modernisation thinking, covenants are believed to induce technological innovation towards more environmentally sound products and processes, although others doubt that and view technological innovation rather as a *conditio sine qua non* for successful application of environmental covenants. An important advantage for the target groups is, finally, that covenants reduce uncertainty about the future course of government policies. Even though public authorities cannot fully bind themselves in this respect (see below), the agreements make drastic changes in the character or content of policies very unlikely.

Nevertheless, the status of environmental covenants is still unclear and controversial. Some covenants are formulated as 'gentlemen's agreements', which means that they are not enforceable by the court. Others have a more binding character, but even then uncertainty exists as to the consequences of this in practice. As negotiations with industry are usually carried out with intermediary organisations, individual members of such organisations are not legally bound by the outcomes of such negotiations. In some sectors and on some issues, moreover, the degree of organisation is low, so that parts of the sector are not represented at all (as in the case of the packaging covenant). This may lead to the paradoxical situation that the most effective way for public authorities to enforce compliance with a covenant, which is in the end an agreement under *civil* law, is with the help of *public* law, namely by introducing customary legislative instruments. Interestingly enough, the introduction of formal legislation is sometimes requested, or at least not opposed, by front-runner industries. The position of public authorities in relation to covenants is thus somewhat ambivalent. With regard to the introduction or the content of future legislation, the government can only bind itself to a 'best effort' obligation to adhere to the covenant. In the last instance, however, the provisions of a covenant can be overruled by public law, as happened for instance in the case of packaging (for more details on these points see van Vliet, 1992; van Acht, 1993; van Buuren, 1993; van de Peppel, 1995).

In addition to problems in the field of implementation and enforcement, another major shortcoming of covenants and the target group approach in general is the closed character of the decision-making process, as clearly demonstrated by the case study of the dairy industry. This limits the opportunities for public and political participation and control, for instance by environmental organisations (see above), but also by representative fora such as parliament.

In late 1995, a broad, evaluative study on the use of covenants was published by the Dutch Court of Audit (Algemene Rekenkamer, 1995). It covered some 154 covenants that had been concluded between the central government and private actors by 1 September 1994. No less than 85 were concerned with environmental issues. Almost half of those (42) were related to energy efficiency and saving and had been concluded by the Ministry of Economic Affairs. The Ministry of VROM was responsible for another 32. The conclusions of the report confirmed some of the drawbacks of covenants referred to above. More than half of all covenants, for instance, turned out to contain insufficient guarantees for effective enforcement. Many of them restricted themselves to mutual commitments to make certain efforts, without concrete, measurable goals and deadlines. This also applied to the environmental covenants concluded by the Ministry of VROM, but it should be noted that this ministry showed a notable improvement on this point after 1990. The report also confirmed the existence of problems with regard to how binding the commitments are on the target group. In several cases, it appeared to be unclear, first, to what extent intermediary organisations indeed covered the entire sector. Second, 'hard' guarantees regarding the commitment of individual firms to the agreements made by their representatives were lacking. In addition to the points mentioned earlier, the Court of Audit criticised the way covenants were selected as the most appropriate policy instrument. Ministries were often hardly able to substantiate why they expected a covenant to be more effective or efficient in a given context than traditional regulatory instruments, and if this outweighed the drawbacks of the former.

CONCLUSION: VOLUNTARY AGREEMENTS AS A STEP TOWARDS ECOLOGICAL MODERNISATION

With regard to the question posed in the title of this paper, it has become clear that, at least as far as the Netherlands is concerned, voluntary agreements cannot easily be equated with deregulation. Deregulation as a goal in itself, as we have seen, has hardly been successful in the Netherlands. Efforts to reduce the regulatory burden on industry in the early 1980s led to little more than a limited wave of reregulation. In the environmental field, these efforts were succeeded by a more encompassing reorientation, centred around the key-word of internalisation. Some aspects of deregulation can indeed be recognised in the internalisation approach, particularly in the attempt to take better account of the needs and wishes of the target groups already in the stage of formulating policy measures.

These attempts, however, were part of a broader strategy, not primarily aimed at restoring industry's competitive strength by creating more room for economic development, but rather at achieving a more positive relationship between economic development and ecological objectives. This would not necessarily lead to a reduction in the number or complexity of rules and, in practice, it indeed hardly did so. The emphasis was rather on the process, the alternative way of making and implementing rules and norms, which was supposed to contribute to a new way of thinking about the problems of economy and ecology. Moreover, in the end, the elements that had been commonly agreed often found their way into 'traditional' legislation or licences. In one respect, the internalisation approach even considerably increased the policy-related burden on firms, in the form of intensive, time-consuming negotiations with the government and the development of internal environmental management and monitoring systems. One might argue that this was the price to be paid for an increased influence on the policy process on the part of industry, especially *vis-à-vis* other private actors.

The philosophy behind the internalisation approach is better associated with the concept of ecological modernisation than with that of deregulation.

Remembering the three basic trends with regard to the interrelationship between state, market and technology postulated by the theory of ecological modernisation (section 2), it can be seen that Dutch target policies aim at transcending the old dichotomy between economy and ecology. The integration of ecological requirements into the design and operation of production processes and the search for positive-sum solutions are explicitly regarded as steps towards the development of ecologically more rational modes of production. The link with ecological modernisation is even more obvious in the political realm. As Le Blansch (1996) stresses, internalisation policy is exemplary for a more preventive, participative and 'reflexive' way of steering. In this sense, the Dutch experience can indeed be regarded as a type of a testing-field for the shifting role of the state in the model of ecological modernisation (cf. Weale, 1992; Spaargaren and Mol, 1992).

NOTES

1 For a general review of Dutch environmental policy, covering both history and organisational features, see Liefferink (1997a).
2 This case study is largely based on an evaluation of the use of covenants in the dairy industry carried out by the Department of Sociology of Wageningen Agricultural University (Chi, 1996).
3 Separate covenants were often concluded for specific problems or sub-sectors within each of these fifteen sectors. This explains the total number of more than 100 environmental covenants that are now in force in the Netherlands.
4 However, Mol (1995) points out that in the early case of the covenant on Volatile Organic Compounds (VOC), negotiated in the period 1985–1988, industry representatives presumed that also the 50 per cent reduction goal itself was negotiable.
5 NZO was founded in 1993 by a fusion of three branch associations (one for the cooperatives, one for the private companies and one for preserved milk producers). Nedsmelt represents 5 cheese melting companies, 3 of which are also members of NZO. The two associations represent 22 of the 25 companies in the dairy sector, including all major ones, and cooperated intensively during the negotiation process.
6 FO-industry is a public organisation set up in 1993 by the Ministry of VROM to serve as a facilitator for covenant negotiations as well as for monitoring and evaluating covenant implementation.
7 On 1 January 1996, a national energy tax for households and small industries was introduced. Large industries were exempted from this tax, however, with the argument of competitiveness. Instead, so-called Long Term Agreements were concluded between the Ministry of Economic Affairs and major industrial branches, including the dairy industry. Also in the discussions about a carbon/energy-tax at the EU level, the Dutch Ministry of Economic Affairs proposed to maintain this approach for larger industries.
8 It is interesting to note that originally a more interactive approach, following largely the Dutch model, had been considered at the EU level for various waste streams, including packaging waste (cf. Chang, 1996). Currently, the possibility of voluntary agreements in environmental policy is again being discussed in Brussels.

9 Cf. for instance the President of the Association of Dutch Christian Employers (*Nederlands Christelijk Werkgeversverbond*, NCW), J.C. Blankert, in *NRC-Handelsblad*, 21 November 1994.

REFERENCES

Aktieprogramma DROM (Deregulering Ruimtelijke Ordening en Milieubeheer), Tweede Kamer 1982–1983, 17931, no. 4.

Algemene Rekenkamer (1995), *Convenanten van het Rijk met bedrijven en instellingen*, Tweede Kamer 1995–1996, 24480, nos 1–2.

Andersen, M.S. (1994) *Governance by Green Taxes*, Manchester: Manchester University Press.

Bressers, J.T.A. (1983) 'Beleidseffectiviteit en waterkwaliteitsbeleid: een bestuurskundig onderzoek', Enschede: Technische Universiteit Twente, unpublished dissertation.

Chang, I. (1996) 'Voluntary agreements in EU environmental policy [working title]', Wageningen: Department of Sociology WAU, unpublished M.Sc. thesis.

Chi, G. (1996) 'The emergence of voluntary agreements in environmental policy making. A case study of the Dutch dairy industry', Wageningen: Department of Sociology WAU, unpublished M.Sc. thesis.

Fischer, H. (1991) 'Harte und Sanfte Chemie am Beispiel der Farben und Lacke', *Wechselwirkung* 48, pp. 15–18.

Gastkemper, H.J. and Klijn, E.H. (1986) 'Ook deregulering leidt tot remise', *Bestuurswetenschappen* 1, pp. 23–38.

Hajer, M.A. (1993) 'The politics of environmental discourse: a study on the acid rain controversy in Great Britain and the Netherlands', University of Oxford, unpublished thesis.

Hanf, K. (1989) 'Deregulation as regulatory reform: the case of environmental policy in the Netherlands', *European Journal of Political Research* 17, pp. 193–207.

Huber, J. (1982) *Die verlorene Unschuld der Ökologie. Neue Technologien und superindustrielle Entwicklung*, Frankfurt am Main: Fischer Verlag.

Huber, J. (1991) 'Ecologische modernisering: weg van schaarste, soberheid en bureaucratie?', in Mol, A.P.J., Spaargaren, G. and Klapwijk, A. (eds), *Technologie en milieubeheer. Tussen sanering en ecologische modernisering*, Den Haag: SDU.

IMP-M (Indicatief Meerjaren Programma Milieubeheer) 1985–1989, Tweede Kamer 1984–1985, 18602, nos 1–2.

IMP-M (Indicatief Meerjaren Programma Milieubeheer) 1986–1990, Tweede Kamer 1985–1986, 19204, nos 1–2.

Jänicke, M. (1979) *Wie das Industriesystem von seinen Mißständen profitiert*, Opladen: Westdeutscher Verlag.

Jänicke, M. (1986) *Staatsversagen. Die Ohnmacht der Politik in der Industriegesellschaft*, München/Zürich: Piper.

Jänicke, M. (1993) 'Über ökologische und politische Modernisierungen', *Zeitschrift für Umweltpolitik und Umweltrecht* 2, pp. 159–175.

Le Blansch, K. (1996) *Milieuzorg in bedrijven. Overheidssturing in het perspectief van de verinnerlijkingsbeleidslijn*, Amsterdam: Thesis Publishers.

Liefferink, D. (1997a) 'The Netherlands: a net exporter of environmental policy concepts', in Andersen, M.S. and Liefferink, D. (eds), *European Environmental Policy: the Pioneers*, Manchester, Manchester University Press.

Liefferink, D. (1997b) 'New instruments in the Netherlands', in Golub, J. (ed.), *New Instruments for Environmental Protection in the EU*, London: Routledge.

Mol, A.P.J. (1995) *The Refinement of Production. Ecological Modernisation Theory and the Chemical Industry*, Utrecht: Van Arkel.

NEPP (National Environmental Policy Plan 1990–1994), Second Chamber, session 1988–1989, 21137, nos 1–2.

NEPP-Plus (National Environmental Policy Plan Plus), Second Chamber, session 1989–1990, 21137, nos 20–21.

Paehlke, R. and Torgerson, D. (eds) (1990) *Managing Leviathan. Environmental Politics and the Administrative State*, London: Belhaven Press.

Schöne, S. (1987) 'Ontwikkeling van doelen en strategie van de energiebeweging: ecologische modernisering', in *Eenheid en verscheidenheid in natuur- en milieudoelstellingen*, Utrecht: RUU, Department of Societal Biology, report of a symposium.

Simonis, U.E. (1989) 'Ecological modernisation of industrial society: three strategic elements', *International Social Science Journal* 121, pp. 347–361.

Spaargaren, G. and Mol, A.P.J. (1992) 'Sociology, environment and modernity. Ecological modernization as a theory of social change', *Society and Natural Resources* 5, pp. 323–344.

Urgentienota Milieuhygiëne, 1971–1972, Tweede Kamer, 1971–1972, 11906, nos 1–2.

Van Acht, R.J.J. (1993) 'Afdwingbare milieuconvenanten?', *Nederlands Juristenblad* 67 (14), pp. 512–517.

Van Buuren, P.J.J. (1993) 'Environmental covenants – possibilities and impossibilities: an administrative lawyer's view', in van Dunné, J.M. (ed.), *Environmental Contracts and Covenants: New Instruments for a Realistic Environmental Policy?*, Lelystad: Koninklijke Vermande, pp. 49–55.

Van de Peppel, R.A. (1995) *Naleving van milieurecht. Toepassing van beleidsinstrumenten op de Nederlandse verfindustrie*, Deventer: Kluwer.

Van Driel, P., Cramer, J., Crone F. *et al.* (eds) (1993) *Ecologische modernisering*, Amsterdam: Wiardi Beckman Stichting.

Van Nispen, F.K.M. and van der Tak, Th. (1984) 'Herregulering', *Beleid en Maatschappij* 12, pp. 367–372.

Van Vliet, L.M. (1992) *Communicatieve besturing van het milieuhandelen van ondernemingen: mogelijkheden en beperkingen*, Delft: Eburon.

VROM (Ministerie van Volkshuisvesting, Ruimtelijke Ordening en Milieubeheer) (1992) *Milieu en industrie. Opzet en achtergrond van het doelgroepenbeleid*, Den Haag: Ministerie van VROM.

VROM (Ministerie van Volkshuisvesting, Ruimtelijke Ordening en Milieubeheer) (1994) *Working with Industry*, The Hague: Ministry of Housing, Spatial Planning and the Environment, Environmental Policy in Action no. 1.

WCED (World Commission on Environment and Development) (1987) *Our Common Future*, Oxford: Oxford University Press.

Weale, A. (1992) *The New Politics of Pollution*, Manchester: Manchester University Press.

Winsemius, P. (1986) *Gast in eigen huis, beschouwingen over milieumanagement*, Alphen aan den Rijn: Samsom H.D. Tjeenk Willink.

Zimmerman, K., Hartje, V. and Ryll, A. (1990) *Ökologische Modernisierung der Produktion. Strukturen und Trends*, Berlin: Sigma.

12

ENVIRONMENTAL MANAGEMENT SYSTEMS IN THE NETHERLANDS

Towards the third generation of environmental licensing?

Frans van der Woerd

INTRODUCTION

In 1989, the Dutch government initiated an action programme to stimulate Environmental Management Systems (EMS) on a voluntary basis. This programme can be considered a forerunner of BS7750, ISO 14000 and the European Eco-Management and Audit Scheme (EMAS). The major goal of the programme was for industrial companies to have EMS in operation by 1995. From 1989 to 1996, a steady development of EMS in large corporations can be observed. While the goals of the 1989 programme have not been achieved, business environmental management has become regular practice in Dutch industry.

When drawing up the action programme, the central government expected that local authorities would change their permitting (use of 'generic licences' or EMS licences) and enforcement (to 'control of EMS') procedures. Business representatives were also in favour of such a development. Deregulation as a reward for business self-regulation was the new motto. No blueprint for a combination of EMS and deregulation was offered at the time. Suggestions were made that by making the management system part of the environmental licence firms would have more freedom in the structuring and execution of their environmental action. Pilot projects were initiated to examine opportunities and problems. Progress with this type of deregulation has been very slow in spite of subsidies and pilot projects. By the end of 1995, eight Dutch companies had a permit based on their EMS. Another eight companies were well advanced in preparing such a permit (Nijenhuis and Aalders, 1995). However, for most companies, licensing and EMS have remained separate.

What are the reasons for the slow pace of the developments? Do companies underestimate the efforts needed to implement a well-functioning EMS? Do local governments refuse to acknowledge EMS results?

Are major bottlenecks in company–government relations an obstacle? This chapter aims to provide answers to these questions. It is based on three years of research into EMS developments. After various theoretical considerations, the chapter uses four case studies to highlight different types of government–company interactions.

The chapter first looks at Dutch EMS developments between 1989 and 1995. It then summarises the history of environmental licensing in the Netherlands. The remainder of the paper focuses on changing company–government relations in response to the introduction of EMS. It begins with a theoretical model, followed by the results from four case studies. To conclude, trends in business and government behaviour are assessed.

Since 1993, EMS developments have become increasingly linked to the evolution of voluntary agreements between public authorities and certain sectors of industry. In fact, pleas for self-regulation of individual companies based on EMS can be seen as the company/local complement to covenants on a sectoral/national level. The preceding chapter by Liefferink and Mol has discussed voluntary agreements in some detail and they are not examined any further in this chapter.

BUSINESS ENVIRONMENTAL MANAGEMENT IN THE NETHERLANDS, 1989–1995

Since the first National Environmental Policy Plan (NEPP) was published in 1989, Dutch environmental policy towards industry has increasingly focused on EMS. In its 1989 White Paper on environmental management systems, the government defined EMS as 'systematic activities of a company aimed at an insight into, control and decrease of its burden on the environment'. This definition was meant to include both technical and organisational activities (VROM, 1989). Additionally, attention was to be paid to all aspects of the product life-cycle (raw materials, production, consumption and waste). Central government and business representatives together elaborated a model for a 'standard EMS'. This EMS includes eight elements, forming a management feedback loop, as shown in Figure 12.1.

When comparing Figure 12.1 to the draft proposal for an international EMS standard, one finds that the two are very similar. ISO 14001 identifies five elements (ISO, 1995):

1 commitment and policy;
2 planning;
3 implementation;
4 measurement and evaluation;
5 review and continuous improvement.

ISO elements 2 to 5 inclusive reflect EMS. Element 1 is essential for management commitment. As ISO 14001 was not available for companies

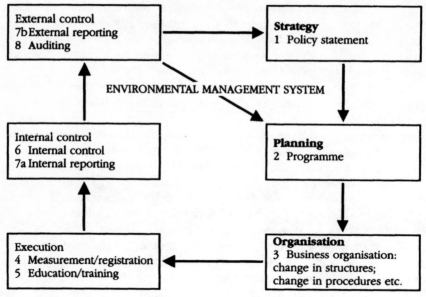

Figure 12.1 Standard EMS
Source: VROM, 1989 (adapted)

during our investigations, we base the following sections on the Dutch standard EMS.

The NEPP's goals were ambitious. By 1995, about 10,000 'large polluters' were to have implemented the standard EMS. For about 250,000 'small polluters', a simplified partial system was to suffice. Both goals were supported by business representatives. We will concentrate on progress with large polluters. Reliable figures for 1995 are not yet available but the indications are that the 1989 targets have not been achieved. Nevertheless, substantial progress has been made. Trends can be found in surveys carried out in 1991 and 1992. The main determining factor for involvement in EMS has been company size. In 1992, 71 per cent of large firms in the Netherlands (amounting to a total of 300) were well advanced in implementing EMS (see Table 12.1). However, only 19 per cent of this select group had implemented all eight elements of the standard EMS (IVA/ KPMG, 1992). In smaller companies, EMS developments have been much slower. EMS involvement also varies between industrial sectors. The chemical industry is clearly ahead, followed by the food-processing industries. Other industries and public utilities lag behind.

Furthermore, qualitative observations reveal that:

- implementation of EMS takes at least 4–6 years;
- integration of environmental aspects in strategic management is often limited;

Table 12.1 EMS implementation amongst 'large polluters'

Company size	1991 (%)	1992 (%)
20–99 employees	8	16
100–499 employees	23	42
> 500 employees	59	71
Average	11	21

Source: IVA/KPMG, 1992

- reporting and feedback of results present substantial bottlenecks.

(Woerd, 1993)

It is clear that it takes much time and effort to develop EMS elements into proper EMS systems. In 1995, a number of studies were published looking at the operation of EMS in the Netherlands. These show that the state of the art can be characterised as 'process optimisation'. Generally, in EMS, waste management and good housekeeping score high, while product life-cycles and prevention score low. EMS are mainly aimed at operational improvements and there is a lack of interest in strategic decisions (Molenkamp, 1995). Le Blansch (1996) has found that the 'internalisation' of environmental values has largely remained superficial.

Over time, the goals of EMS may evolve. To assess EMS aspirations, models describing the different stages of EMS have been developed. Usually, three stages are distinguished in these models:

- reacting to government rules;
- anticipating future government rules;
- preparations of own rules.

Petulla has labelled these stages 'crisis-oriented, cost-oriented and enlightened environmental management' (Petulla, 1987). In this paper we distinguish between the following stages, as identified by Erasmus University (1990):

1 *'Inspection'*: reactive, obligation, non-systematic.
2 *'Total compliance'*: pre active, cost optimalization, partial systematic.
3 *'Total integration'*: proactive, prevention, complete systematic.

In accordance with the figures discussed above, it is no surprise to find that bigger companies are generally heading for 'total compliance'. The majority of smaller companies are still in a stage of 'inspection'. In general, proactive EMS are rarely found.

Preparing its own environmental rules would effectively mean that a company is self-regulating. However, several authors have pointed out that complete business self-regulation in environmental matters is not conceivable (van Driel, 1989; SER, 1994). Self-regulation never works in a vacuum, it is not a 'do-it-yourself regulation'. Different groups in society will influence the contents and rules of business self-regulation, directly and indirectly. Seeing the environment as a public good, governments have the task of establishing and monitoring overall policy goals. This task cannot be transferred to individual companies. Therefore, practical discussions should not concern complete self-regulation, but 'institutionalised self-regulation', i.e. a shift in responsibilities towards companies. We call this governmental re-regulation, applied in combination with EMS-based self-regulation.

Before discussing re-regulation and new generations of environmental permits in more detail, the next section will give an overview over the Dutch regulatory framework and its development between 1970 and 1995.

THREE GENERATIONS OF ENVIRONMENTAL LICENSING

As mentioned earlier, the complementary evolution of EMS implementation and less detailed environmental licences was expected. The Environment Ministry stated in 1989 that 'Environmental management will not replace existing regulations but will be complementary' (VROM, 1989). The regulatory framework was not changed for EMS, which meant that permitting authorities had to find new ways of using the existing framework. This section describes the regulatory framework and its implementation. The goal is not to give a general picture but to stress the elements that determine the scope of local/regional authorities.

From the early stages of Dutch environmental policy in 1970, the command-and-control approach has been dominant. The issue of permits was the most important policy instrument. Economic and communicative instruments were supplementary. Effective command-and-control regulation assumes a well-functioning 'regulation chain'. Figure 12.2 presents its elements.

As in EMS, the backbone of the regulation chain is a feedback-cycle. However, government regulation is more complicated than EMS, because:

- The implementation of strategies to comply with permit requirements lies outside the influence of public authorities. Without specific measures, business policies remain a 'black box' for the authorities. EMS offer opportunities to open up the box.
- Government control is a necessary complement to permits. Enforcement was neglected in the Netherlands until 1985. Although much effort has been put into improving the situation since, the 1995 situation is not deemed optimal. EMS may improve the situation.

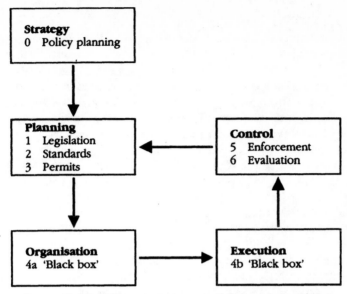

Figure 12.2 Regulatory instruments and the regulation chain
Source: VROM, 1985 (adapted)

- Three levels of government share responsibilities for policy planning: national, regional and local authorities. There is no formal hierarchy between the three levels. Therefore, coordination mechanisms are needed.

In the Netherlands, national government is responsible for legislation and nationwide targets. Also a limited number of national emission standards exists. According to the Environmental Policy Act (1993), integral permits are issued by the provinces (for large companies) and municipalities (for small companies). For discharges to water, a separate permit is issued by the Water Boards.

An additional problem is that Dutch legislation consists of framework laws: targets are not included in the law. Therefore, provinces, municipalities and the Water Boards are relatively free in choosing the contents of their permits. This freedom is motivated by the notion that legislation should be adaptable to local/regional circumstances. In principle, EMS can be included in a permit. However, the legislation does not explicitly mention this possibility. The inclusion of EMS in a permit depends on the initiatives of regional/local authorities. Legislation is one factor among many others explaining the decisions of permitting authorities. These authorities act semi-autonomously (Aalders, 1987).

Between 1970 and 1990, two types of licensing developed:

- strict command-and-control licences, still dominant in municipalities;
- negotiated licences, dominant in the provinces.

Strict command-and-control licensing means detailed permits, based on technical prescriptions. These prescriptions are aimed at the operational level. Negotiated licences start with consultation between company and authority about environmental problems, possibilities and priorities. Next, joint conclusions are documented in a less detailed permit, based on emission targets. This type of licensing became dominant for large polluters in the 1980s. Negotiations typically took place at the tactical level, with the objective of identifying win-win solutions (van der Tak, 1988).

EMS-based self-regulation offers possibilities for government re-regulation, as mentioned before. Companies implementing integrated, dynamic policies could be given more responsibilities. In that case, EMS can open up the 'black box' and stimulate government enforcement. However, questions arise as to the result of this for actual licensing and enforcement. We call this the third generation of licensing.

The precise contents of the third generation licences is not yet clear. Important elements will be quantitative targets for a time period of at least four years for strategic environmental aspects. EMS will be important in monitoring progress. Finally, authorities will agree about environmental priorities and coordinate enforcement. These issues are further explored in the next two sections, first through a theoretical framework, followed by a look at 'learning-by-doing' in this new and complicated field.

THEORETICAL PERSPECTIVES OF COMPANY–AUTHORITY INTERACTIONS

This section presents a framework for the analysis of interactions between a company and an authority in concerted actions to achieve environmental goals. The framework has been derived from the Standard EMS (Figure 12.1), the Regulation Chain (Figure 12.2), supplemented by network models (policy research) and stakeholder models (business research). The model operates within the following framework conditions:

- Both companies and authorities are 'open systems', having contacts with different actors in society. It is too simple to discuss only company–authority interactions (the bipolar model).
- In the Netherlands, licences are issued by provinces and municipalities, not by central government. Hence, local governments have a specific position in the model. Through the granting of licences, they can directly influence business behaviour.
- The licensing authority interacts with other authorities. For example, national government plays a role as policy-maker, while independent Water Boards are responsible for water pollution.
- Interactions and influences flow in both directions. Company and authority meet in an 'arena for bargaining'. No single actor is omnipotent:

companies have bargaining power because of their detailed knowledge of production processes, while local authorities have bargaining power because of legislation.

Our framework is presented in Figure 12.3. The company side looks very much like stakeholder models in management literature (Kast, 1980). The authority side is much more simplified. The 'arena for bargaining' is the core of the framework.

The framework in Figure 12.3 is too simple for a detailed analysis of interactions. Therefore, the interactions have been specified in Figure 12.4. The arena for bargaining couples the feedback loops of EMS (a planning-and-control model) and of government policy-making (a command-and-control model). This picture uses common characteristics of information and feedback in business environmental management and in government policies. Figure 12.4 combines the core elements of Figures 12.1 and 12.2. The result is an extended EMS feedback loop:

permit requirements ⇒ EMS planning ⇒ internal control ⇒ enforcement.

From a practical point of view, Figure 12.4 shows that two fields of interaction require special attention:

- liaisons between permits and company planning;
- liaisons between company control and enforcement.

Flows of information, as well as time, are essential for effective interactions. In this respect, the concept of a 'learning curve' must be introduced. In the 'arena for bargaining', the learning curve has three aspects. Firstly, experience has shown that implementing EMS in a company

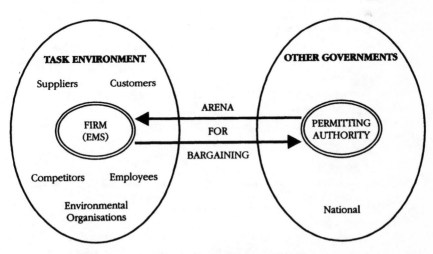

Figure 12.3 Framework for company–authority interactions

205

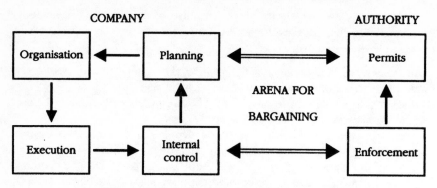

Figure 12.4 Feedback loops and interactions

takes four to six years. Secondly, government agencies face the complicated task changing their traditional technical/juridical approaches into more administrative/management approaches. Thirdly, both companies and governments must find ways to dovetail their environmental actions. Taking into account these three points, it is clear that both companies and authorities need learning time before EMS can be coupled effectively to re-regulated government actions. Table 12.2 shows a possible evolution in interactions. This table combines the previously mentioned three stages of EMS with the three generations of licensing.

What can we expect from learning-by-doing? Will EMS bring about a breakthrough? The foregoing theories point at possibilities and bottlenecks to be expected. Here, we will discuss the following four hypotheses.

1 Interactions based on a fully functioning EMS (= total integration) are possible, when all organisational provisions for internal feedback in a firm are implemented. This expectation imposes heavy demands on EMS implementation. To close the feedback loop inside the company is far from easy. As shown before, very few Dutch companies have reached that stage as yet.
2 Interactions that reap the fruits of an active EMS are possible, if all authorities concerned have integrated their permit and enforcement actions. The second expectation is the mirror image of the first one.

Table 12.2 Classification of business–government interactions

	Company	Authority
No EMS	Inspection	Command and control
	Total compliance	Consultation
EMS	Total integration	Generic licensing

206

Government actions and priorities should be predictable for a company. Therefore, in order to create unanimity, authorities should, in the first place, establish mutual relationships.

3 During the introduction of EMS in a firm, the number of prescriptions in a permit will not decrease, but technical prescriptions will be replaced by organisational prescriptions. It is yet unclear which EMS elements can and should be included in environmental permits. As EMS are quite complicated as systems, the paradox of EMS prescription by governments may occur, whereby permits become even more complicated than before. Recently, researchers advised not to include the complete EMS in a permit (Nijenhuis and Aalders, 1995). They argued that by doing so, the licensing process would become very cumbersome and inflexible.

4 Feedback of high-quality information by companies is of greater importance for the development of new interactions than formal prescriptions in a permit. We assume that both firm and authority strive for a win-win game. Their ultimate benefits are reductions in uncertainty. To improve the effectiveness of interactions, the best thing a company can do is provide timely, high-quality information, irrespective of formal demands.

COMPANY–AUTHORITY INTERACTIONS IN FOUR CASE STUDIES

In this section, we will analyse actual changes in government–business interactions after the introduction of EMS, based on four case studies. The four case studies present a clear picture of Dutch experience up to 1994. By that time, only eight companies had completed all steps towards EMS permits. The cases point at possibilities and bottlenecks but it will become clear that it is too early to draw statistical conclusions. After a description of the case studies, answers to the preceding hypotheses will be formulated.

Company 'A' is a plastics and metalware factory (producing for the professional market), which employs 3,100 people. Its environmental problems relate mainly to (chlorinated) solvents, heavy metals and metal plating. In 1988, an EMS was initiated. Between 1988 and 1994, discharges of heavy metals to water and emissions of solvents to air have decreased substantially. Good housekeeping and soil sanitation continue to be matters of concern. In 1994, management decided to apply for BS 7750 by 1996. In 1990, the company got the first Dutch permit based on EMS, issued by the municipality. In 1995, the separate permit for water pollution still does not include EMS elements. As a result, two regimes in licensing run parallel.

Company 'B' is a producer of inorganic chemicals with 180 employees. Company B produces intermediates, sold on professional markets. Its products are not poisonous. Process emissions relate to dust, noise and water and there is soil contamination. Management decided in 1990 fully to implement EMS by 1995. In 1994, both company and authorities considered emission levels as satisfactory. However, environmental procedures and internal feedback are far from complete. In 1993, the procedure to obtain new permits based on EMS started. In spite of the preceding negotiations, it took one year to get a 'hybrid' provincial permit, considered unsatisfactory by both company and province. Reasons for the dissatisfaction were a lot of misunderstandings and information provided by the company which was too limited. The separate permit for water pollution met with similar, but narrower, problems.

Company 'C' develops and produces telephones and tracer systems, mainly for professional use. It employs 380 persons. Being an electronic product industry, environmental problems concentrate on solvent emissions from painting and cleaning. In 1990, a new factory was built, with soil contamination being effectively prevented on the new site. In 1992, a start was made with EMS. Because of company reorganisation, progress has been limited so far. Actual planning anticipated full EMS implementation by 1996. In 1993, the municipality initiated a new type of enforcement within the limits of the existing (non-EMS) permit of 1991. The company produced an annual environmental report. Results were considered promising by all parties involved. The municipality plans to expand this experiment to other companies.

Company 'D' is a tanker cleaning company in one of the Dutch seaports. Chemical waste from ships is processed by 50 employees. Not surprisingly, environmental impacts are big and diverse (waste water, waste, risks, but also emissions to air and noise). After conflicts between the company and environmental authorities, in 1991, the newly appointed management decided to implement an EMS by 1993. The branch association provides the company with standard EMS and yearly audits. By 1994, implementation with regard to daily operations was almost finished. However, preventative actions were still in their infancy. This company has three permits: standard environment and water permits plus a permit for the processing of chemical waste. After two years of negotiations, in 1992, new permits based on EMS were requested. The three authorities involved diverged completely: the municipality agreed, the water board agreed half-heartedly and the national government (promoter of EMS!) refused. After two years of discussions, this unsatisfactory situation continues.

Having introduced the four case studies, we analyse our findings with respect to the four hypotheses discussed previously. The first statement assumed that new interactions are only possible after the full implementation of EMS. This appears not to be true. In all cases, new interactions began before the full implementation of EMS. However, this approach involves a certain amount of risk: extra efforts are demanded (case C) and temporal set-backs occur because of communication failures (case B). On the whole, the process can be considered as learning by experience.

The second hypothesis assumed that fruitful interactions are only possible when government policies are coordinated. This is not the case. Agencies are not closely tied, they have room to manoeuvre independently. In order to avoid conflicts, informal contacts are necessary. In most cases, these types of contact exist in the Netherlands. If not, useless efforts and mutual frustrations arise (case D). These findings confirm earlier studies (Aalders, 1987; van der Tak, 1988).

The third hypothesis was that EMS permits would not be simpler than traditional permits. This hypothesis also fails to hold. EMS-based permits can be very simple indeed. The minimum length is two pages, prescribing annual programmes/reports, regular audits and a few emission limits (case A): i.e. the elements of the extended feedback loop (Figure 12.4). For the time being, hybrid forms of permits prevail (case B + D). They are longer than in case A, but more simple than previous permits.

Finally, we discuss the hypothesis that flows of information are more important than formal permits. This has been fully confirmed by our research. Regular, timely and reliable information is the core of the new interactions. All case studies focus on information. After the implementation of EMS, the annual programme/report will play an important role in interactions.

As an important conclusion we can say that win-win solutions are possible, but there is no free lunch. Win-win in practice means reductions in uncertainty. To accomplish this, both company and authority have to invest in time and money. In the short run they have to learn the game; in the long run, they have to produce and discuss plans and reports.

CONCLUSIONS

The case studies have shown that re-regulation coupled to EMS is possible. Annual business programmes/reports, based on the EMS planning and reporting cycle, play a crucial role in this approach. A start can be made before EMS has been fully implemented. However, this brings along risks for companies and agencies as misunderstandings, temporal set-backs and frustrations are likely to occur.

In all cases, information flows are the core of the new interactions. Well-developed information channels are essential, both between firms and government agencies and between agencies. Information is the mortar for mutual confidence. In the absence of information channels and high-quality information, useless efforts and frustrations prevail. The 'arena for bargaining' (Figures 12.3 and 12.4) has not changed as such. Rather, its use will change and become more intensive.

Full development of new interactions takes a number of years. All participants have to exert themselves in order to change their own organisation. Additionally, new interactions require their own learning time. Overall, the process can be regarded as a complicated and time-consuming one, which will take a number of years. It is the duty of the authorities to establish quantitative environmental targets. EMS can be used to accomplish these targets.

Before entering the procedure for 'third generation licensing', both companies and authorities have to be aware of possibilities and requirements. In the first place, they have to ask themselves:

• Do we want new types of interactions?
• Are we able to meet the necessary requirements?

Companies need a solid basis, to be able to draft an ambitious environmental programme and to demonstrate its accomplishment. Agencies need clear policy guidelines, to be able to draft quantitative targets for individual companies, to integrate enforcement and to coordinate actions with other agencies. Only when these requirements have been met, is it wise to start negotiations for licence renewal.

The overall conclusion is that dovetailing EMS and deregulation is possible. However, it is not a panacea for all environmental and enforcement problems. The process towards 'third generation licences' is a supercritical one: if one actor does not cooperate, efforts may be in vain. In 1996, most Dutch companies and authorities are not ready for these developments. In the course of time, learning curves and examples from other companies may change this situation.

REFERENCES

Aalders, M.V.C. (1987) *Regeltoepassing in de ambtelijke praktijk van Hinderwet en Bouwtoezichtafdeling*, Groningen: Wolters-Noordhoff.
Driel, M. van (1989) *Zelfregulierung*, Deventer: Kluwer.
EUR, Erasmus University Rotterdam 'Le Manageur' (1990) *Milieubedrijfsvoering, problemen en perspectieven*, Rotterdam.
ISO (1995) Draft international standard for EMS (ISO 14001), Geneva.
IVA/KPMG (1992) Bedrijfsinterne milieuzorgsystemen-tussenevaluatie 1992, Tilburg/Den Haag.
Kast, F.E. (1980) 'Scanning the future environment: social indicators', *Californian Management Review*, 23 (1), p. 24.
Le Blansch, C.G. (1996) 'Milieuzorg in bedrijven – overheidssturing in het perspectief van de verinnerlijkingsbeleidslijn', unpublished Ph.D., Utrecht University.
Molenkamp, G.C. (1995) 'De verzakelijking van het milieu: onomkeerbare ontwikkelingen in het bedrijfsleven', Den Haag: KPMG.
NEPP (1989) *National Environmental Policy Plan*, Den Haag: VROM.
Nijenhuis, C.T. and Aalders, M.V.C. (1995) *Naar een flexibele vergunning. Koppeling van het milieuzorgsysteem aan de milieuvergunning*, Amsterdam: Centrum voor Milieurecht.
Petulla, J.M. (1987) 'Environmental management in industry', *Journal of Professional Issues in Engineering*, 113 (2), pp. 67–75.

SER, Dutch Social-Economic Council (1994) *Advies over het Nationaal Milieube-leidsplan 2*, Den Haag.

Tak, Th. van der (1988) 'Vergunning verleend', unpublished Ph.D. thesis, Delft University, Delft.

VROM (1985) *IMP Milieubeheer 1986–1990*, Den Haag.

VROM (1989) Notitie bedrijfsinterne milieuzorg, Den Haag.

Woerd, F. van der (1991) *De manager en het milieu*, Amsterdam: IVM-VU.

Woerd, F. van der (1993) 'Five years of Dutch experience with environmental management systems', paper for the symposium Environmentally Sound Products with Clean Technologies, Budapest, 23–27 August.

13

EPILOGUE

Ute Collier

The aim of this book has been to examine different facets of the deregulation concept from an environmental perspective – a perspective which is frequently absent in the discussions surrounding the issue. Undoubtedly, deregulation, be it in terms of market liberalisation or regulatory reform, has important implications for environmental protection. The evidence to date is that there are both costs and benefits. There can be little doubt that the operation of unfettered markets will not provide a sufficient level of environmental protection. Thus, if deregulation and liberalisation are taken too far, without accompanying compensatory measures, additional environmental damage can occur. At the same time, a certain amount of economic liberalisation and the use of alternative regulatory instruments can have environmental advantages.

Several chapters of this volume have argued that economic deregulation needs to be accompanied by a regulatory framework to ensure environmental protection, providing, for example, for access to environmental information, ensuring investments in energy efficiency and renewable energies, or safeguarding water quality. There is scope for using market-based instruments, including self-regulation, but these need to be carefully evaluated, both in terms of their environmental effectiveness and in terms of their impact on different industrial sectors or on firms of different sizes.

To some extent, environmental protection in Europe is at a cross-roads. Since the 1970s numerous environmental regulations have been adopted. Some improvements in environmental quality have occurred, but the future holds many more problems, such as climate change and continued resource exploitation. Some of the problems have been caused by government economic policies themselves. A serious pursuit of the sustainable development concept will require many changes and certain deregulatory aspects may be included, be it through the use of environmental taxes or by opening up energy markets to independent renewable energy generators.

Furthermore, sustainable development also implies changes in the concept of governance, with a greater involvement of citizens and industry and stronger cooperation between private and public actors. However, this approach can bring its own problems, for example with industry employing its vast resources to influence decision-making in their own interests. Overall, there needs to be a careful mix between deregulation and intervention, between cooperation and decision-making in the interest of the public good, and a balance must be struck between minimising costs for industry and ensuring a high quality environment for society.

How likely is such a balance in the current political climate? Once more it is instructive to look at the UK, which, after having been in the forefront of liberalisation moves in the EU, has also been leading the movement towards deregulation. In June 1994, the UK conservative government established a Deregulation Task Force under the Deregulation and Contracting Out Act, consisting almost entirely of business representatives, with only one member from a charitable organisation and no involvement of independent analysts.

The Task Force was directly responsible for the introduction of a requirement on government departments to accompany all new regulatory proposals with risk assessments and compliance cost assessment, apparently using highly controversial methodologies (ENDS, 1996a). Environmental policy was one of the focal areas of the Task Force's work, although its membership contained no environmental experts, and it asserted that environmental legislation is a 'major burden' on business (Deregulation Task Force, 1996). As this volume has discussed at length, this assertion is open to debate and the Task Force's expertise in this area has to be questioned considering its one-sided membership.

It is thus not surprising to find that in the environmental field, the Task Force was blamed for putting enforcement at risk (ENDS, 1995). In response to one of the recommendations of the Task Force, the Department of the Environment has instructed the newly established Environment Agency to draw up a code of enforcement practice which gives polluters extra opportunities to object to proposed measures (ENDS, 1996b). Effectively, the procedures involve extra bureaucracy, which may inhibit already overburdened regulators taking action and can hardly be called deregulatory. This is especially paradoxical as, in its latest report, the Task Force is particularly critical of the Department of the Environment's proposals to implement the EU packaging waste Directive because of the administrative costs it is likely to entail. The main problem is that the Task Force appears to start its deliberations from the premise that as soon as environmental measures impose costs on business, they should be abandoned. As it is not clear how many 'free lunches' are actually available, this stance poses a considerable threat to environmental protection in the UK.

As in the case of liberalisation, few other governments in the EU are going as far down the deregulation track as the UK did under the Conservatives who were determined to assert their deregulation ideology at the EU level as well. They claimed to be 'changing the regulatory climate in Europe' (Her Majesty's Government, 1996). Indeed, it appears that deregulation and subsidiarity together are having a major impact on policy developments in the EU, not only in the environmental field. The number of legislative proposals has fallen dramatically since 1992. This is not necessarily a bad sign as such, as one could argue that legislation has become so extensive that it might be a time for consolidation and for addressing the well-documented 'implementation gap'. However, one cannot help feeling some unease at proposals amending existing legislative measures, such as the revisions to the drinking water Directive, which would result in a weakening of various standards, without any compensatory measures.

Much rhetoric continues about the preference for economic instruments but to date little progress has been made. Economic instruments might be preferable from an environmental point of view, as a means to tackle the compliance problem. However, it is not clear whether they would be efficient enough to make the use of regulatory standards obsolete. Also, the cost imposed on business can be just as high as with regulatory standards. Hence, politically, economic instruments are often no easier to agree on than regulatory standards. In fact, sometimes it can be more difficult as the costs, in the case of taxes and charges can be more transparent.

Meanwhile, self-regulatory instruments continue to be emphasised as the preferred alternative by business, but as experience from the Netherlands demonstrates, there are uncertainties as to whether they can really work. Naturally, it is preferable that environmental protection measures minimise the costs to businesses and consumers. However, unfortunately it is not easy to assess costs and benefits, in particular with problems such as climate change, which are subject to large scientific uncertainties. Current deregulation initiatives put far too much emphasis on short-term costs, while often ignoring long-term benefits.

As the turn of the century approaches, deregulation, competition and globalisation are dominating the political agendas. Long gone seem the days when environmental issues hit the headlines, yet the Rio Summit only lies a few years back. Sustainable development seems to have almost slipped into oblivion in what some have called the 'era of competition' (Petrella, 1995). Some voices of dissent have been raised. In 1995, at the same time as the Molitor Report hit the headlines, a report by the Lisbon Group (a group of independent experts from ten countries), entitled 'The Limits to Competitiveness', was published (Petrella, 1995). The report warns about the short-sightedness and narrow-mindedness of the competition ideology. It argues that both liberalisation and deregulation are based on rules of conflict. As an alternative, the report promotes cooperation between governments, businesses and private actors for the achievement of a sustainable, equal world. To date, there is limited evidence that such cooperation is forthcoming to tackle environmental problems. Deregulation, competition and liberalisation are being pursued as if they were an end in itself, rather than a means to an end and one of several alternative possibilities. In the EU, the philosophy behind the idea of European integration is almost exclusively economic and other issues such as environmental protection continue to be marginalised.

REFERENCES

Deregulation Task Force (1996) *Report 1995/96*, London: Deregulation Unit.

ENDS (1995) 'Enforcement practices caught in deregulation drive', *ENDS Report* no. 248, September 1995, pp. 28–29.

ENDS (1996a)'Task Force wants more radical deregulation', *ENDS Report* no. 260, September 1996, pp. 29–30.

ENDS (1996b)'Deregulation of enforcement puts Agency at risk', *ENDS Report* no. 252, January 1996, p. 3.

Her Majesty's Government (1996) *The Government's response to the Deregulation Task Force Report 1996*, London: Deregulation Unit.

Petrella, R. (ed) (1995) *I Limiti della Competività*, Rome: Manifestolibri.

INDEX

abatement costs issue 134
access to information: an alternative
  regulatory mechanism 58–9; CEE
  countries 79; in a deregulated
  environment 62–3; goals of 57–9;
  organisation of in a deregulated
  environment 64–71; public–private
  participation 67; treatment of trade
  secrets 69–71
acid rain, and the CEGB 98, 99
'actio populensis'-type procedural rights 83
adaptive regulation 16–17, 50
administration: costs high for SMEs 149,
  150, 150; procedures 68, 76;
  redefinition of functions 59–60;
  reforms 68
'agency capture' problem 99
Agenda 21 17
agriculture: environmental taxation of
  inputs 141; pollution by 132
air pollution 108; control of and
  abatement costs 134; industrial, by
  SMEs 148; regulation in CEE countries
  79; and transport 116
air transportation, growth in 116
anticipation, of future environmental
  considerations 173, 174
authorisation procedures 76

bacteria, breaking down pollutants 132
balancing tests, for trade secrets 70–1
'best available' technology 45, 116
biochemical oxygen demand (BOD) 132
Brickman, R., Jasanoff, S. and Ilgen, T. 157
Britain–Germany alliance, pressing for
  deregulation 160–1
British Gas 99
Brundtland Report 171, 186

Brusco, S., Bertossi, P. and Cottica, A.
  147, 154
BS 7750 60, 171–2, 207
bulk goods, transport of 117
business and environmental
  deregulation 165–80

Cairncross, F. 36
California, innovations in environmental
  policy 157–8
capital, availability to SMEs 150
capital investment, availability post-
  privatisation 140
carbon/energy tax 13, 14, 97, 111
carrying capacity, maintenance of 16,
  25–6, 27–9, 32, 37; need for
  government intervention in the
  market 29–31
CCGTs *see* combined-cycle gas turbines
CEE countries *see* Central and Eastern
  European countries
CEGB *see* Central Electricity Generating
  Board (UK)
Central and Eastern European countries
  75; attitudes to deregulation 76;
  environmental provisions 77–81
Central Electricity Generating Board
  (UK) 98, 99
chemical companies, and environmental
  regulation 156
chemical industry, USA, cost-benefit
  risk-based approach to regulation 158
chemicals/dangerous substances,
  provisions reasonable, CEE countries
  79
CIT *see* countries in economic transition
Clean Air Act (USA) 158–9
climate change 107

reductions 110; energy sector liberalisation 96, 99–106; environmental technology market 153; lacking market for heat 103; narrow reading of trade secrets 70; Regional Electricity Companies (RECs) selling efficiency 106
UK Central Electricity Board 7
UK Environmental Information Regulation 64
UN/ECE draft Convention on Access to Environmental Information 57
UNCED, Agenda 21 57
UNICE 149; and the SID concept 173–4
UNICE Regulatory Report 19–20, 162, 165, 167–9, 170; calls for action programme 168–9; finds EU regulatory regime anti-competitive and growth-impeding 167–8
USA 5, 161; environmental deregulation in 157–60; Environmental Protection Agency 159; President's Council on Sustainable Development 159; tradable permits 136–7

victim compensation 136
voluntary action, reasons for 177
voluntary agreements 20, 137–8, 199; as a form of deregulation? 181–97; and internalisation 184–7, 194–5; may foster innovation and efficiency 138;

as a step towards ecological modernisation 194–5

waste disposal industry, lax enforcement in 150
waste furnace gas 107
waste management: regulation, CEE countries 80; and the smaller firm 154–5
water meters, lacking in UK 139
water pollution 131–3; charges 140–1, 142; instruments for control of 19, 133–8; non-point sources 132, 141; point sources 132
water quality: CEE countries 80–1; and pollution 133; UK, objectives set by Environment Agency 139
water sector: current environmental regulation in Europe 138–41; market and regulatory failure in 131–44; privatisation and liberalisation 19, 139–40, 142
water supply 131–2; domestic and industrial use 132
The White House 158
'win-win' situations 12, 35, 186, 209; and sustainability 29–30
World Business Council for Sustainable Development 28
World Industry Council for the Environment (WICE) 28, 172, 175